The Insects

Url Lanham

The Insects

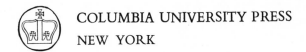

COLUMBIA UNIVERSITY PRESS
NEW YORK

Printed in the United States of America

10 9 8 7 6 5

Dedicated to the memory of the entomologist
T. D. A. Cockerell, a friend, who wrote as an
introduction to his college textbook of zoology:

To live, to work, to grow, to love,
Shall earth below or heaven above
Ask more of thee?
Thus holding fast the golden thread
Which joins the living and the dead
Through all eternity!

Contents

The line drawings were made by Caroline Lanham, the wife of the author. The photographs were made by the author.

Part one

In perspective

I

The place of insects in nature

The insect world is the most pleasingly intricate of all animal worlds. Insects are small, but under the microscope their sculptured armour reveals varied complexity of structure, which in turn reflects myriad ways of life lived out in the tens of thousands of kinds of miniature habitats of the world. The variations in structure and mode of life are without practical limit, for there are more kinds of insects in existence than of all other animals taken together—about three fourths of a million species are known—and it is said that less than one half, perhaps one tenth or even only one twentieth, of the species now alive have been discovered and described. The exploration of this myricolored biological jungle has scarcely begun and will never be ended.

Insects live on the land or in fresh water; scarcely any live in the sea. One can tow a net through the blue water of the ocean far from land for hours, capturing thousands of small animals, and not one will be an insect. But sweep a net for a few minutes through the vegetation of a sunny meadow, and the net swarms with a host of small animals, all of them insects, except for a few spiders.

Insects are thus creatures of the land, but their distant ancestors, which gave rise to them by a long series of evolutionary changes, were marine, and to understand the insects it is necessary to start far back in their history, with animals that lived in the sea.

Of the approximately twenty major groups (phyla) of animals that exist, all have representatives in the sea. Of these, only nine have produced animals able to live on the land. Furthermore, most of these nine groups have made only a partial conquest, in that they can be active only in environments where the surroundings are moist, as in the soil or under the bark of fallen logs, where water is not lost from the body by evaporation. Snails (of the phylum Mollusca), for example, are found even in semidesert regions, and at first one might think them truly creatures of the dry land. But most of the time these snails lie dormant, a waterproof seal stretched across the shell opening, and crawl about only in the rain or at night, when dew lies on the ground.

Only two of these nine phyla have produced animals that are able to live indefinitely in truly dry environments, that is, where the air is dry and where special adaptations are necessary to keep the animal from drying out. These are the phylum Chordata—which contains man and other such familiar vertebrates as the other mammals and the birds—and the phylum Arthropoda—which contains the insects as well as a number of other groups. Both the chordates and the arthropods had their origin in the sea. The first by way of fishes and amphibians produced the vertebrate animals that were to succeed on the land, and some of the ancient ocean-dwelling arthropods produced centipede-like forms, which in turn gave rise to the insects.

In order to understand the structure of an arthropod it is necessary to consider two other phyla that figure in the evolutionary history of these animals. These are the annelid worms (phylum Annelida) and Peripatus (phylum Onychophora). The annelids probably are the group that invented the basic body plan that characterizes both them and the arthropods, and the annelids probably gave rise to the arthropods. Both are segmented animals, that is, they are built on a module plan in which the body is made up of more or less similar units (segments) laid end to end. Some of these segmented animals grow by adding new segments; in others the final number is already blocked out in the embryo.

The majority of the annelid worms live in the sea—and some of these are active or glowingly colored animals—but the most familiar example is the earthworm. Each of the many short segments of this animal repeats to some extent the internal structure of its neighbors. Externally the earthworm is nearly featureless, the short bristles that help it move through its burrow being so small as to be nearly invisible. Some of the marine annelids have soft flaps on the segments that help in swimming, but none have good walking or swimming legs. It is the development of a pair of such legs, or appendages, on each segment that makes for the important difference between annelids and arthropods. Another significant difference—one that contributes to the perfection of these appendages—is the usually hardened, armor-like skin of the arthropods, which contrasts with the usually soft skin of the worms.

The gap between the annelids and the arthropods is bridged to some extent by the obscure onychophorans. The fifty species of these caterpillar-like, richly colored animals live in the southern continents and northward into the tropics of Central America and the Orient. Their soft, velvety skin is not waterproof, so that they are restricted to such humid environments as mossy rock crevices and rotten logs. On the head are a pair of small crystalline eyes and a pair of antennae. These slow-moving creatures defend themselves by squirting out from the head ribbons of sticky substance, and they also entangle their prey with it.

Among the characteristics of onychophorans that have been held to place them on a line of evolution between annelids and arthropods are their numerous pairs of stubby, soft legs, which are perhaps intermediate between the flaps, or parapodia, of the annelid and the armored, jointed legs of the arthropod. The breathing apparatus of the onychophorans—a system of tubes that conduct air deep into the body—and the heart are like those of land-dwelling arthropods. The body muscles are, on the contrary, like those of the annelids in their slowness of movement and in their microscopic structure. Only a small set of muscles that operates the biting jaws (which are the only hardened structures of the animal) is the same as those of the quick-moving arthropods.

A main objection to the hypothesis that the onychophorans form in any way an evolutionary link between annelids and arthropods is that they are terrestrial, whereas any such missing link should be marine, since this evolutionary transition must have taken place in the sea. Biologists were therefore much interested when fossil onychophorans were discovered in marine rocks of the Cambrian age, the Burgess Shales of British Columbia.

If a soft, worm-like animal resembling an onychophoran were to acquire protective armor by hardening the skin, the advantage would be obvious: predators fitted to cope only with soft-bodied prey would, until they themselves evolved new offensive weapons, be

FIG. I. *Distribution of the Onychophora and the major groups of arthropods through geological time.*

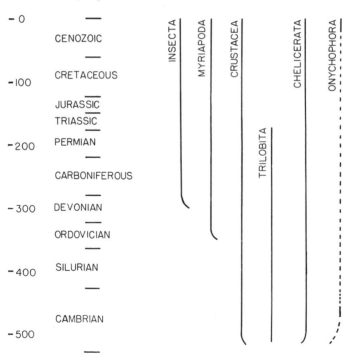

thwarted by the appearance of such creatures. The value of such a defense in the ancient Cambrian seas is shown by the fact that a group of true arthropods with armored skins, the trilobites, were then the most common and diversified of the larger animals, to the extent that this oldest geologic period for which there is a good fossil record has been called the Age of Trilobites. Although the trilobites are extinct, they must be reckoned as remarkably successful animals, since they were in existence for well over 500 million years, with a longer history than that of all the vertebrate animals.

The impression given by the neatly sculptured body of the trilobite in stone is one of armored strength. The name comes from the division of the body into three lengthwise lobes: a central thicker one, which contained all or most of the internal organs, and a pair of thin lateral flanges which covered the legs and gills (Fig. 2). A wide head and tailpiece fitted snugly against the body trunk. When the animal coiled up, the vulnerable legs and gills on the underside were completely protected.

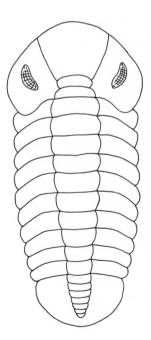

FIG. 2.   *Trilobite. The antennae and the appendages are not shown in this diagram of a fossil.*

Trilobites were marine animals and in their long history did not invade rivers or land. They ranged in size from one-fourth inch to over two feet. In their prime, in the Cambrian, they were among the larger animals, but by the time they were approaching extinction at the end of the Paleozoic some 200 million years ago they were dwarfed by a host of larger aquatic predators, including the strong-jawed fishes.

Judging from the structure of their legs, trilobites were able to crawl on the bottom; although some may have been swimmers, they could not have been very good ones. The legs of trilobites were slender structures, with a branch at the base bearing a feathery gill. Since the legs were, like the rest of the body, covered with an armored skin, they were jointed to allow for movement, with narrow bands of flexible skin dividing the leg into a number of rigid segments. Segmented or jointed legs (or appendages, as they are often called, for they were frequently transformed into structures that have nothing to do with locomotion) are typical of the whole arthropod phylum and are responsible for its name (from the Greek, *arthron*, joint, and *pous, podon*, foot).

The ancient trilobites had evolved sense organs on the head that are still found in most living arthropods: a pair of large faceted (compound) eyes, and a pair of long feelers, or antennae.

The initial advantage of armor is, obviously, that of giving mechanical protection, especially against predators. However, the consequence of this seemingly restricted invention has been far-reaching and has provided the basis for the remarkable success achieved by the arthropods in all the main environments of the world.

Of perhaps most general significance, the armor provides firm anchors and points of attachment for the muscles. That is, it becomes a skeleton, which, being on the outside, is called an exoskeleton. Muscular action now becomes quick, powerful, and precise, and, along with the acquisition of a skeleton, the muscles change from the slow-moving, smooth type found in such invertebrate animals as the annelid worms and molluscs to the striated type found in arthropods and the vertebrate animals.

Since the legs become also incased in armor, they can become longer and, as efficient levers, make it possible for their possessors to run or swim faster than can be done with short, fleshy legs. More than this, such hardened legs, or appendages, can be converted to uses other than locomotion, and the arthropods have developed this potentiality to a remarkable degree, the legs being transformed into devices used for such purposes as digging, chewing, stinging, grasping prey, and even constructing shelters.

Finally, the armored skin has been drawn out into planes that have become the wings of insects. Insects are the only invertebrate animals that can fly, and it is the ability to fly which has contributed most to making the insects, and hence the arthropod phylum, the most varied and abundant of land animals.

The ancient trilobites took little evolutionary advantage of the opportunities offered by the exoskeleton. They were well armored for protection, and their long legs made it possible for them to move faster on solid footing than their annelid ancestors did. But there was little in the way of modification of the legs for the specialized uses that we find in other arthropod groups. All the many pairs of legs of a trilobite were pretty much alike, with the exception that those around the mouth had expanded bases that could help crush food and get it into the mouth. All the legs were used in walking. The other groups of arthropods, by contrast, may delegate the function of walking to a few pairs of appendages, that of feeding to other pairs, that of seizing of prey to another, and so on. Different ways of dividing responsibilities among the pairs of appendages characterize the different groups of arthropods, and it is these variations that provide most of the subject matter of a discussion of the arthropod groups which surround and lead up to the insects.

A summary classification of the arthropods needed as a reference is given in the following list.

Phylum Arthropoda
  Subphylum Trilobita
    Class Trilobita. The trilobites are extinct, having lived from Cambrian to Permian times; all were marine.

Subphylum Chelicerata
 Class Merostomata—includes the extinct eurypterids (sea scorpions) and
 the living horseshoe crabs (xiphosurans). Live in coastal waters, salt and
 brackish.
 Class Arachnida—scorpions, spiders, mites, and so forth. These are
 terrestrial, except that some mites have invaded the water.
Subphylum Mandibulata
 Class Crustacea—shrimps, lobsters, crayfish, barnacles, pill bugs, and
 so forth. Most crustaceans live in the sea and in fresh water, but a
 few are terrestrial.
 Class Insecta. The insects, or six-legged arthropods, live in all habitats
 except the waters below the surface of the open sea.
 (The remaining classes are for convenience placed in a more or less
 informal group called the Myriapoda.)
 Class Diplopoda—millipedes. These live on the land, usually in damp
 and hidden habitats.
 Class Chilopoda—centipedes. All are terrestrial carnivores.
 Class Pauropoda. The pauropods are very small, little-seen animals,
 all terrestrial, that live in concealed habitats.
 Class Symphyla. In appearance and habits somewhat like the preceding,
 the symphylans are of special interest because they seem to connect the
 six-legged insects with their myriapod ancestors.

The living classes of arthropods fall into two main divisions.
One consists of the chelicerates, in which the first pair of legs has
been modified into a pair of pincers, the chelicerae. In the other
division are the mandibulates, in which the first pair of legs has
either been converted into a pair of antennae, or is discarded, and
in which the second pair of legs has been transformed into a pair
of jaws.

The chelicera is an adaptation for a predatory, carnivorous life, and
the chelicerate animals were from the beginning committed to it, so
that all living forms are—like the spiders—strictly carnivores, except
that some of the mites have become adapted for drinking plant juices.

Some of the extinct chelicerates were among the largest and most
powerful carnivores of their time. The eurypterids, or water scorpions,
which flourished in Devonian times (about 300 million years ago) and
became extinct in the Permian (about 100 million years later), were

as much as six feet long, and some had spiny, powerful chelicerae that must have dealt effectively with a variety of prey (Fig. 3). Some had a pair of long, paddle-shaped legs that presumably made them fairly good swimmers. Eurypterids lived in brackish and fresh waters near the coast, in company with primitive, small fishes, which were the first vertebrate animals. For a time the eurypterids dominated these vertebrates in point of size and strength and undoubtedly preyed upon them, but gigantic fishes appeared by the end of the Devonian, putting to an end once and for all the early challenge to supremacy made by the arthropods.

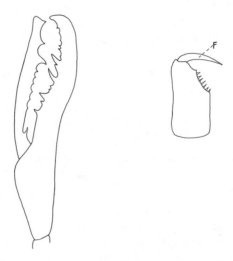

FIG. 3. *Chelicerae of a giant eurypterid (left) and of a spider (right). The eurypterid chelicera is ten inches long. The spider chelicera is equipped with a hollow, movable fang (F) through which venom is injected into the prey.*

The xiphosurans, another ancient branch of the chelicerates, lived at the same time and in much the same places as the eurypterids. They have survived with little change to the present as the horseshoe crabs of the coasts of the western Atlantic and the tropical Orient. Horseshoe crabs are not as agile as were the eurypterids, but they have specialized in armor protection, most of the body being covered by a smooth, vaulted hemisphere of exoskeleton so tough that a good can opener is proper equipment for starting a dissection of one of them. This shield may be a foot across. Fitting behind it is a movable plate, armed on the edge with spines and covering what corresponds to the abdomen, followed by a long spike-like tail.

In spite of the primitive, trilobite-like appearance of the horseshoe crab, the appendages concealed under the armored shield show the animal to be comparatively specialized. Behind the weak chelicerae, fitted only for picking up such prey as worms and small molluscs, are five pairs of walking legs. These legs surround the mouth and resemble the front legs of the trilobite in that the base is thickened so as to help handle the food. Several pairs following, on the abdomen, are so changed as to be almost unrecognizable. Each leg is branched, with the upper branch carrying hundreds of small flat leaves that provide a large respiratory surface for absorbing oxygen from the water—these are the gills—and a lower branch shaped into a flat plate that protects the gills.

Scorpions are eurypterid-like animals that have become adapted for life on land. Although they are among the largest of the land arthropods, their maximum size of eight inches hardly matches that of their gigantic extinct marine relatives. In front of the four pairs of walking legs is a pair that has been modified into large pincers used in holding prey, and most anterior of all are the chelicerae, also used in handling prey. Although the long abdomen is without recognizable appendages, except for a pair of comb-like structures of uncertain function, the embryonic scorpion has short leg-like structures with gill pockets developing near them. In the adult these become four pairs of respiratory organs called book lungs, which are much like the gills of the horseshoe crab but are concealed in a moist pocket within the body where they absorb oxygen from the air.

Scorpions are given added offensive power by the large sting on the tip of the abdomen. This is thrust forward with precision to inject into the prey a venom produced by a large gland in the bulb of the sting.

The best known of the chelicerate animals are the spiders. They live mostly on insects, a dependable, ubiquitous source of food. These they subdue with the aid of the chelicerae, which have been converted into poison fangs (Fig. 3). They walk on four pairs of legs. Two leg-like structures behind the fangs, called pedipalps, are often held out like antennae and probably serve in part as sense organs. In the male the tips of the pedipalps are modified into curious mating organs: the male deposits a drop of sperm, blots it up with the tips of the pedipalps, then inserts one of these structures into the genital opening on the abdomen of the female. Unlike the insects and most other arthropods, spiders do not have compound eyes but have several (as many as eight) simple eyes—minute, glassy hemispheres set in the front of the cephalothorax.

Spiders owe their success above all to the manufacture and ingenious use of silk. The silk is made as a viscous plastic in large glands that nearly fill the globular abdomen. This liquid plastic is squirted out through a set of spigots, or spinnerets, at the tip of the abdomen, and it usually becomes solid when stretched and exposed to air. These spigots are complicated, and the openings can be adjusted by muscular control so that a variety of silk threads can be produced on demand: fine simple strands, thick multiple cable, bands beaded with sticky drops of unsolidified silk, or flat bands used to ensheath the struggling prey.

Spiders put the silk to use early in life. Soon after hatching, the young spider usually launches itself into the world by means of an aerial balloon voyage. The tiny spider climbs up on to a high place, then plays out a very fine line of silk. This is caught in air currents and, when long enough, pulls the spider away from its perch. These balloons are very effective: even the updrafts caused by warm reading lamps in a still room are strong enough to carry the spider into the air. At certain times of the year drifting spiders may fill the air, and, when they descend on open fields, the balloons may form filmy sheets of silk acres in extent. When these sheets are torn by the wind, the sky may be filled with these glistening scraps of silk, called gossamer.

Many spiders hunt their prey cat-fashion, by stealth, with a final rush or leap. This is the method used by the tarantulas, large primitive spiders which are the bulkiest of all dry-land arthropods. A South

American tarantula has a body three and a half inches long and weighs nearly three ounces; the legs spread over nine inches. Tarantulas make docile pets, rarely using their poison fangs to bite. On extreme provocation, a tarantula brings the hind pair of legs up over the abdomen and, with a motion almost too quick to see, strokes its furry back. The observer sees a light cloud of dust float up from the tarantula, but the significance of this does not dawn on him until he begins to gag and choke, as a result of inhaling the particles of broken poisonous body hairs. A small mammal attacking a tarantula must find the experience a memorable one.

Wolf spiders (gray or brown spiders that run swiftly over the ground) and jumping spiders are abundant and diverse groups that do not use webs to trap their prey. Wolf spiders often hunt at night, and the collector finds them by using a headlamp, which makes the eyes of even the smallest ones shine like brilliant green or yellow jewels. The jumping spiders are active in daylight and are the most gaily colored of the spiders.

Silk is used throughout the life of many spiders as an anchor line. As it walks, the spider plays out a silk thread, which sticks at intervals to the surface. If we allow one of the jumping spiders to walk out to the tip of our finger, then bring another finger up close, it may jump across and, as it does so, leave the strand of silk bridging the gap. If we force the spider to jump without a landing place, it does not fall to the ground but squirts out the silk line, brakes to a stop, and climbs back up the line to its initial resting place. The dragline is primarily a device that helps to keep the spider from getting lost.

The most familiar use of silk by spiders is the construction of webs used to trap insects. The bolas spider goes about the trapping business aggressively. By a series of complicated maneuvers it produces a sticky ball of silk on the end of a short line, with a front leg whirls the ball around, lets it fly at the prey, then hauls the captured insect in. The arachnologists have, luckily, obtained photographs to document their story of this mode of predation.

The insect-trapping webs of many spiders are erratic tangles spun in corners or over vegetation. Others are precisely constructed flat, circular sheets with a large and efficient trapping area, sometimes with a central opaque geometric design that may serve to attract insects. The very complicated behavior patterns necessary to make these specialized webs are inherited, just as the complex body structure is inherited. Spiders can be classified in part by web structure, as well as by body structure. When the spider is fed insects that have been injected with drugs known to cause behavorial abnormalities in other animals, the web is abnormal and can be studied to gauge the strength and qualitative effects of the drug.

Finally, silk also is used by the spider toward the end of its life (*see* the classic E. B. White, *Charlotte's Web*) to construct an elaborate waterproof shelter for the eggs and young.

Mites are diminutive relatives of spiders that perhaps exceed in numbers of individuals all other arthropods on the land. Almost any handful of soil is likely to contain tens or hundreds of them. Although most live hidden in the soil, some live in exposed situations for at least part of their lives. In early spring large scarlet, velvety mites are seen crawling slowly over the ground. The chigger is a newly hatched mite that bores through the human skin, then injects a poison that causes a great itching welt that lasts for a week or longer. The more it is scratched, the more the welt humps up to surround and protect the mite. In safety the mite drinks predigested tissues for a few days, then drops off, gorged with a meal that provides it with enough nutrient to grow to maturity.

The mites that feed on warm-blooded animals can recognize sources of heat. This ability for thermodetection was strikingly shown by some mites that had flourished in a neglected laboratory mouse colony so much that they covered one of the cages with a gray, crawling film. A finger held half an inch above them caused the mites to form a small mound as they climbed up over each other to reach the warmth that promised a source of food. That it was not the sense of smell that stimulated them was shown when they tried to reach a flame held over them. When a cigarette lighter was held a few feet below the cage, the mites crowded to the warm draft of air and eagerly poured in a cascade into the flame.

Mites that feed on mammals are able to transmit microbial diseases from one individual to another. Dangerous microorganisms carried by mites are the rickettsias, which in size lie between the minute viruses and the larger true bacteria. In World War II the Allies landed troops on the little island of Owi, lying near the equator north of New Guinea, since it offered a good site for an airfield and was free of enemy troops. Also it had no native inhabitants, an ominous indication of the disaster that was to strike the American troops who landed there. The island was infested with mites that carried a rickettsia causing scrub typhus, and a large percentage of the Americans were made severely ill by the disease. The Owi Island strain of scrub typhus turned out to be rarely fatal, but elsewhere there were many fatalities from the disease.

Ticks essentially are large mites (Fig. 4). These flat, leathery animals, all bloodsucking parasites of other animals, have beaks furnished with backward-directed hooks that make it almost impossible to pull one off once it has begun feeding. Like some of the parasitic mites, the ticks may infect their victim with rickettsias. Rocky Mountain spotted fever is one of these tick-borne diseases.

The chelicerate animals are arthropods that were early committed to a carnivorous mode of life. Since their prey is usually only crushed,

FIG. 4.   *Tick.*

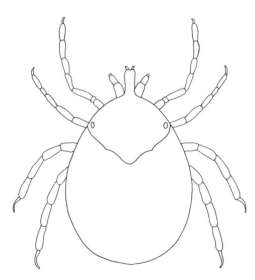

or perhaps pulled to pieces, then predigested and consumed as a liquid, no elaborate set of mouthparts was needed for processing the food. Thus, only the first pair of appendages (chelicerae) and sometimes the second pair (pincers of the scorpion) became specialized for feeding. The following four pairs were devoted to walking, the rest discarded. Thus, the early commitment of the chelicerates to a carnivorous existence resulted in the early restriction of the number of pairs of appendages, and this produced evolutionary inflexibility in adapting to other modes of life which require complex mouthparts, composed of three or more pairs of highly modified appendages, to handle other types of food.

By contrast, the mandibulate group of arthropods always had available a large number of pairs of appendages for use in one or another evolutionary venture; or at least they did so until the appearance of the insects, who retained only three pairs of walking legs, discarding the rest.

The oldest of the mandibulate arthropods are the crustaceans. Fossils found in the Cambrian rocks along with the trilobites are those of already complex and well-specialized crustaceans, so that the group must have arisen in pre-Cambrian times. Their ancestors probably were arthropods that resembled the trilobites in having appendages not specialized for any function other than crawling but that differed from trilobites in not emphasizing armor protection. The Crustacea were founded as a group—and a highly successful one—by the specialization of several anterior pairs of legs for food handling and of others for swimming. Although in their later history many of the crustaceans have become slow-moving armored animals, or sedentary or even parasitic ones, the great majority are marvelously adapted for life in the waters of the open sea, where they swim and drift throughout their lives. Apparently none of the other arthropod groups are thus adapted but are dependent at some stage of their existence on the bottom or the surface.

So abundant are the crustaceans in the ocean that they will nibble away the carcass and thus strip the skeleton of a large whale in a matter of hours; and in turn many of the whales grow to their full size of hundreds of tons in a few years by feeding on small crustaceans strained out of the water. Their abundance and their diversity have earned for the Crustacea the title of "the insects of the sea."

Although the identity of the appendages concerned in the evolution of the arthropod groups is a matter of controversy among the specialists who try to determine this from the evidence of paleontology, embryology, and comparative anatomy, one interpretation seems to be dominant and is undeniably useful. This is the interpretation owed largely to R. E. Snodgrass, the American morphologist.

According to this scheme, the antennae are not derived from a pair of legs but from a pair of sensitive palps of the kind found on the head of many annelid worms. The first pair of legs, which in the spiders were the chelicerae, in the crustaceans is transformed into a second pair of antennae, which differs basically in structure from the first pair of true antennae. In the crustaceans the second pair of legs becomes a pair of hard, biting jaws, capable of grinding up even refractory plant material. At least the third and fourth pairs also are specialized for handling food, holding it in place and manipulating it for the most efficient operation of the jaws; they also are fitted with sense organs that help evaluate the food.

Crustaceans are an extensive and diverse group, and the appendages behind the mouthparts are modified in many different ways in the different subgroups (Fig. 5). Sometimes there are a number of pairs of walking legs, followed by others used in swimming; or there may be no walking legs but a large number used for swimming only. In

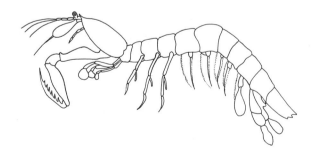

FIG. 5. *Crustacean. This diagram illustrates the large number and diversity of appendages in some representatives of this group. (After Smith.)*

the completely sedentary barnacles, the legs have been transformed into feathery structures that reach out into the water, strain out food particles, and bring them back into the mouth.

From their home in the sea, the crustaceans have gone with success into fresh water, where they range in size from such microscopic, perpetually swimming animals as *Daphnia*, much used by aquarium keepers to feed small fish, to the heavy, bottom-dwelling crayfish. From the fresh-water crustaceans there have evolved permanent land-dwelling forms, including the familiar sow bugs and pill bugs, which are found under stones or wood. These have become proficient air breathers but are restricted to places with high humidity.

The first true arthropods have never been found and remain creatures of our imagination. These animals should have been segmented and slender, with a pair of legs, each like the next, on each segment. The head would have carried no true appendages, only a pair of antennae and compound eyes. The animal would have lived in the sea. The trilobites, the chelicerates, and the crustaceans all could have evolved from a type like this, by the flattening of the body into an armored shield or by the alteration of the walking legs into structures with new functions.

Our theory requires that a group of undifferentiated arthropods like these invaded the shoreline early in the history of life on the continents, perhaps in Ordovician or Silurian times (400 or 350 million years ago) and here gave rise to the remaining classes of living arthropods, including the insects.

Probably the living animals that look most like these early invaders are the centipedes, and these agree with our ideal concept except in one important respect: the head is more complex, having fused to it the three following segments with their appendages, which are used as mouthparts (Fig. 6). In spite of this major evolutionary advance, the centipedes are probably fairly good representatives of a very early stage in the evolution of the mandibulate arthropods of the land.

The four classes of many-legged arthropods that live on land can be called, collectively, the myriapods. Two of these classes—the pauropods and the symphylans—are small, obscure creatures, usually known only to the specialist. The other two—centipedes and

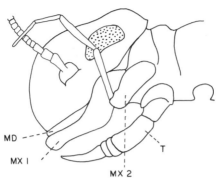

FIG. 6. *Head of a centipede, side view. MD, mandible; MX 1, maxilla; MX 2, second maxilla; T, poison fang, or toxignath, which is the modified first leg. (After Snodgrass.)*

millipedes—are larger, more conspicuous, and better known. Usually almost anyone can distinguish between these two at a glance, like the entomologist's wife who shouted up from the basement, "Hey, there's a centimeter down here—no—I mean a millimeter!"

All of the centipedes are active carnivores. They are able to run swiftly on their confusingly numerous pairs of legs. A way to get increased speed is to lengthen the legs, and this has been done in the swift house centipede (*Scutigera*), which has also simplified the problem

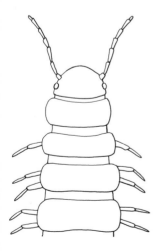

FIG. 7. *The head and first few body segments of a millipede.*

of coordination by reducing the number down to fifteen pairs. Even then, the legs tend to get in each other's way, and groups of legs have to be moved almost in unison, like the oars on a galley, which surely must be a redundancy, one corrected in the six-legged arthropods, the insects.

Centipedes overpower their prey with venom, which is injected by a pair of poison fangs formed from the first pair of legs (Fig. 6). Some of the larger centipedes can pierce the skin of a man with these fangs and inflict a painful bite.

Millipedes are herbivores. They are slow-moving, driving themselves through the litter under stones and logs with powerful thrusts of their numerous short legs. Body segments are fused together in

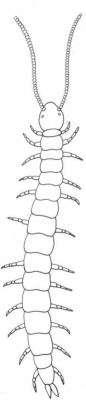

FIG. 8.   *Symphylan.* (*After Snodgrass.*)

pairs, so that there seem to be four legs per segment (Fig. 7). As groups of legs move in unison, waves of motion pass rhythmically down the length of the body. Some of the millipedes defend themselves with armor, which may be a skin-hardening deposit of lime. Others secrete obnoxious or poisonous substances, including hydrogen cyanide. A species living in New Guinea is said to be able to squirt for some distance a poison that can permanently blind an attacker.

The diminutive pauropods and symphylans have fewer legs than most of the other myriapods, having the number standardized at between nine and fifteen pairs (Fig. 8). They are found in the soil or on the damp underside of fallen wood. Some are slow-moving and ornately armored; others are relatively agile and centipede-like. Mostly they interest the student of the origin of insects, for the symphylans, in particular, show many links between the myriapods and the insects.

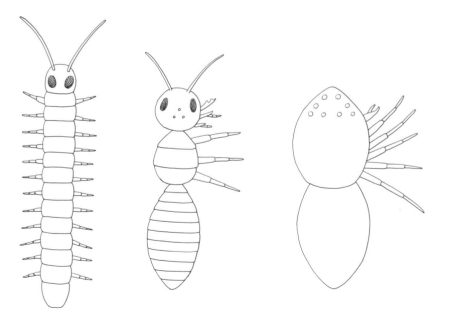

FIG. 9. *Body plans and specialization of appendages of the hypothetical ancestral arthropod (left), an insect (center), and a spider (right).*

Essentially what would be required to change a myriapod more or less like a symphylan into an insect would be to discard all but three pairs of the walking legs, keeping the legless segments as the body region which houses most of the internal organs and using the region which has the legs mostly to house muscles that move the legs. The myriapod, composed of head and trunk, thus would become transformed into an animal with the typical insect structure: a head, with sense organs and feeding appendages; a thorax, with organs of locomotion; and an abdomen.

2

The diversity of insects

Mapping out the diversity of the insect world is in many ways analogous to the work of the topographer, but, where the latter brings in from the field only notes and measurements and photographs, the entomologist brings into the laboratory specimens of insects from the biological terrain. So vast is the insect world that in no museum is there an adequate representation of the kinds of insects in existence. Some of the great and long-established museums of Europe, such as the British Museum and the Paris Museum, come closest to completeness.

For adequate description these museum specimens must be studied minutely in the laboratory. From long experience the student of insects has learned that the best way to handle the fragile specimens is to mount them on slender steel pins (Fig. 10). Thus, the typical insect collection consists of rows of pinned specimens in trays and boxes (Figs. 11 and 12).

Some of the characteristics used in describing insects of some groups are, however, recorded from the living animal. Tape recordings of the songs of sound-producing insects are useful, and often the investigator must keep colonies of the living animal to get details of behavior that he needs for classification. But for the preliminary mapping of insects, which is far from complete, the entomologist relies mostly on

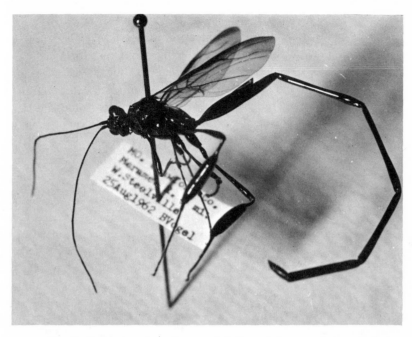

FIG. 10. *The fragile insect specimen (here a pelecinid wasp) is mounted on a pin for easy handling. The printed locality label, about one-half inch long, is made by a photoreduction process.*

the remarkably intricate external structure of the dead museum specimen.

The specimens are arranged, like the books on the shelves of a library, according to a classification that makes it possible to store or find any of several hundred thousand kinds of insects. This classification is not wholly arbitrary but is a reflection of an underlying order of animate nature that has intrigued man, with his urge to classify and explain, from earliest times.

The system for classifying insects that has been developed during the past two hundred years has become extensive and complex. In the mid-eighteenth century Linnaeus listed and briefly characterized all the species of animals, including about two thousand species of insects, known to him; all this was in a single volume of ordinary size.

The situation has since changed a good deal. Marlowe's Doctor Faustus, after selling his soul to the Devil, asked first for a wife and second for books, including one wherein he "might see all plants, herbs, and trees that grow upon the earth." A modern Faust interested in insects would get a rather good bargain—about two thousand large volumes (allowing one page per species) plus an annual supplement of fourteen volumes for the new species described each year.

Although the literature that organizes the diversity of insects in a

FIG. 11. *In preliminary sorting, museum specimens are blocked out into major groups, here the Apoidea, or bees.*

system of classification is in many languages, the formal or scientific names of the species of insects are the same throughout the world. At the time of Linnaeus, who initiated the currently used system of naming plants and animals, the international language of scholars was Latin. Since then, by general agreement, the names of species and of higher categories are Latinized. The international structure of insect classification is at present held together by the *Zoological Record*,

FIG. 12. *These specimens, in the University of Colorado Museum, have been sorted into an arrangement where they can be used in the final phase of research on the composition of part of the insect fauna of a region in the Western United States.*

published in Great Britain, a modest yearly index of the world literature on animal classification in which new species and alterations of
the system of classification are recorded.

In technical works the scientific name of the species is usually
followed by the last name (sometimes abbreviated) of the person who
first described the species and proposed its name. This serves to
differentiate between two names that are alike and gives some idea
as to where the original description might lie in the sea of literature
on systematic entomology. The scientific name (but not that of the
author of the species) is italicized to denote that it is the formal name
of the species.

When a biologist describes a new species, he designates one specimen
as a "type." This is supposed to be preserved in perpetuity as a kind
of international yardstick to indicate exactly what is meant by the
name that is used for the species. The type specimen is the final court
of appeal in questions of the identity of the species. In some museums
there are many thousands of type specimens. They are obviously of
great permanent value, and in time of war the types of the great
museums are moved to safety together with art treasures and irreplace-
able books.

The name of a species of insect is of two words like the name of a
person, that is, it is a binomial. However, the last name comes first,
the more general or generic term being followed by the specific term.
No two generic names of animals are supposed to be alike. When
a biologist finds it necessary to coin a new generic name, he consults
the volumes that list the tens of thousands of generic names that have
already been used to make certain that his is a new one. The specific
term can be used more than once (although not within the same
genus), so that such useful ones as *alba* (white), *coloradensis* (of Colorado),
borealis (northern), *minuta* (small), or *smithi* (referring to a collector
or biologist named Smith) have been used hundreds of times. Since
no specific term can be used more than once in combination with a
given generic term, the name of every species of insect is unique.

When Linnaeus undertook to name and describe all the species of
animals, he regarded each as a "kind" of animal, in the sense used by
Noah when he took aboard a pair of each. Implicit in this is the

concept that two individuals of a species are enough alike that they could be related as parents and offspring, or cousins, and so on; that is, that the members of a species are held together by bonds of heredity. This is basically the modern idea of what is meant by the term species, although with a better understanding of the way in which hereditary materials (genes) behave in populations of this kind it has become obvious that the species is usually in a state of change, that in a local area the individuals of the species may be different from those elsewhere, and that these local races may become so isolated and eventually so distinct from the parent species that they form a new entity no longer able to mix its hereditary material with that of its ancestral species.

The modern concept is, then, that the species is dynamic rather than static. This makes the task of the systematist interesting, even if difficult. These changes have been assumed to take place over millennia rather than decades, but, with close observation of species in nature, it has become apparent that local races, at least, and perhaps species, arise and become distinctive within the lifetime of a man.

Sometimes the local or geographic race is designated by adding a third Latinized name to the binomial, which gives it the rank of subspecies.

A species may be defined as the most extensive population within which genetic exchange—that is, mating—occurs. But, like any definition, this one is not airtight; nature presents situations that are not covered by it. For example, a species may be able to live only in habitats that are widely scattered, separated perhaps by hundreds of miles, so that individuals of one isolated population never have a chance to interbreed with individuals of another. Therefore, such populations are genetically isolated and should qualify as species; but it might happen that, if members of one population were carried to the home ground of another, they would freely mix and interbreed with the residents. If one then changes the definition to read "within which genetic changes could occur," he finds, when he tries to cross supposedly different species, that in more or less artificial conditions they do cross (although he finds no evidence of this in nature), sometimes producing healthy offspring (at least so far as he can tell by observing them in the laboratory), sometimes abnormal offspring, or sometimes none at all. In any event, the procedures necessary to demonstrate genetic isolation are far too time-consuming to be carried out for all of the thousands of species that would have to be dealt with by the pioneer systematist who is mapping out new terrain.

It is a tribute to the analytic ability of the trained mind that the limits of the species as drawn up by the systematist who is thoroughly familiar with the appearance of his animals and who has had long experience examining tens of thousands of specimens usually match very well the underlying genetic boundaries when these have been worked out by long and costly procedures. Perhaps the best working definition is, then, that a species is what a good systematist calls a species.

The systematist has in the back of his mind the idea that the species is the interbreeding population. He has no such yardstick to use in establishing the higher categories—genera, families, orders—necessary to file the some three quarters of a million species of insects. This is not to say, however, that there is no order in nature above the level of species and that any classification in which species are grouped into a hierarchy of categories is purely arbitrary.

Higher categories exist in nature because evolution does not proceed at an even rate; small quantitative changes may have surprising and overwhelmingly important qualitative significance, thus quickly inaugurating whole epochs of evolutionary history that produce groups of animals like none that went before.

Probably the basic circumstance that produces the unpredictable turns and accelerations of evolution is that the environment is constantly changing, not only with the seasonal changes and year-to-year fluctuations but over long stretches of thousands and millions of years. For example, the continents for hundreds of millions of years were barren environments without food for animals, but, when green plants at last became established on them, the environment had changed, so far as animals were concerned, in a fundamental way. The variations among shore-dwelling animals that better fitted some for staying out of the water were before of little, or even negative, value to their possessors, but now they became extraordinarily useful. Their possessors now became, not misfits, but pioneers in one of the most promising realms of the earth, and they founded categories of animals which perfected air-breathing devices, together with whole complexes of coordinated characters built around this adaptation for life on land.

The pioneers sign their own death warrants, of course, for the settlers that stand on their shoulders, exploiting their modest though fundamental inventions, become in turn better adapted for the new life and in competition do not spare their predecessors.

Evolution thus goes by fits and starts, with extensive plateaus in which the great, swift revolutionary advances are slowly exploited in unending detail. This, together with the extermination by new competitors or predators, accounts for the unevenness of the organic world, for the fact that there is not a continuous spectrum of life, or, in other words, for the existence of higher categories.

It is obvious that the criteria for setting up these categories are intangibles and that where the lines are drawn and what rank is given the groups are matters of individual taste. Just as historians group the epochs and periods of human history in different ways, so the systematists, who deal with the results of evolutionary history, set up varied classifications. There are, however, ancient and deep cleavage lines that are relatively unmistakable, so that in spite of the differences in the classifications set up by various authors, there are fundamental similarities. It is not so much a matter of difference of opinion as to the existence of the cleavages, and hence of the major categories, as it is a matter of what rank to give the category that produces the diversity of classifications one sees in books on insects.

Higher categories used within the class Insecta are:

Order
Family
Genus

The family name is always based upon a generic term and is formed by adding "-idae" to its stem. Thus, Chrysomelidae from *Chrysomela*, or Apidae from *Apis*. A family may contain one genus or many. The name of the order is not based upon a generic name but is a coined Latinized term that expresses some attribute of the order. When Linnaeus classified the insects, he based the major divisions upon wing structure and gave the orders names ending in "-ptera" (from the Greek, *pteron*, wing, as in Neuroptera, Lepidoptera, Hymenoptera, and so forth). This tradition has generally been followed since, but there are exceptions, as Odonata (from the Greek, *odon*, tooth, referring to the mouthparts).

These categories are in practice too few, and the system is expanded by use of the prefixes "sub-" or "super-." An expanded system of categories would then be:

Order
Suborder
Superfamily
Family
Subfamily
Genus
Subgenus

The superfamily name ends in "-oidea," the subfamily in "-inae." The subgeneric name is a Latinized word often, but not necessarily, based on a generic name, as *Cryptandrena* (based on *Andrena*). A generic name may be dropped to subgeneric rank, or the subgenus made a full genus, without change in spelling. Other more informal categories, such as tribe, division, or series, are sometimes added to expand the system further.

The honeybee, *Apis mellifera*, would be classified as follows:

Order Hymenoptera
Suborder Clistogastra
Superfamily Apoidea
Family Apidae
Subfamily Apinae
Genus *Apis*

What follows is a brief resumé of the orders of insects that will give the reader a background and reference points for the subsequent chapters on the structure and the ecology of insects. The orders are taken up individually and in detail in later chapters.

The orders are grouped into two major divisions, or subclasses.

They are quite unequal in size: the subclass Apterygota, one of primitive, wingless insects, has in it four orders and less than one percent of the living species of insects; the subclass Pterygota, or winged insects, contains the remaining twenty-two orders. The Apterygota are primitively wingless; that is, they are living representatives of the stage in insect evolution when wings had not yet evolved. Although many species of the subclass Pterygota are in fact wingless, they are believed to have lost their wings secondarily and to be descended from winged ancestors.

With one or two exceptions, all the orders are represented on all continents.

1. Order Protura. Very small, pale, eyeless, soft-bodied. Of the four orders of Apterygota (the others are the Collembola, Diplura, and Thysanura) these are the most aberrant. They are unique among all insects in lacking antennae and in adding body segments as they grow. Like most of their primitive, wingless relatives they live in the soil, under bark, and similar dark, humid, protected situations. The deficiency in antennae is made up by holding the front legs out in front of the head. They are rarely seen and are not well known.

2. Order Collembola—springtails and snow fleas. Very small, sometimes black or even brightly colored, distinguished by their ability to leap. They jump, not with the legs, but with a terminal appendage that is carried folded under the abdomen. They are by far the most abundant insects in the soil, where they occur by the hundreds of thousands of individuals per acre. Sometimes collembollans swarm on the surface of ponds, where they rest on the surface film, and on snow that has had pollen and other organic detritus sifted on it.

3. Order Diplura. Small or medium-sized, rarely large; pallid, eyeless. Their diagnostic feature is a terminal pair of long filaments, although in some the filaments are modified into a pair of forceps used to catch prey. They are not often seen and are little known.

4. Order Thysanura—silverfish and firebrats. Small or medium-sized; sometimes covered with silvery scales; active, some able to leap; large-eyed. The three long terminal filaments are diagnostic.

Of the apterygotes, these are best adapted for active life in open, dry situations. Some are common household pests.

5. Order Orthoptera—grasshoppers, cockroaches, and relatives. Usually large; front wings narrow, flat, slightly thickened, covering the wide, fragile, fan-like hind wings when the wings are in repose. This is a very diverse group, by many authorities divided into a number of distinct orders. The feeding appendages are of a generalized type, from which those of many of the following orders could have been derived, and include a pair of biting mandibles. A very important group, some are major crop pests and others are important herbivores of natural communities by reason of their large size and great numbers. The major divisions of the order are five.

A. Suborder Blattaria—cockroaches. These swiftly running, flattened orthopterans are essentially tropical, but a few species are widespread in dwellings, and some live in the northern hardwood forests.

B. Suborder Saltatoria—grasshoppers (locusts) and crickets. These orthopterans are unable to run but have the hind legs specialized for leaping. Although best represented in the tropics, there are many species in temperate woods and grasslands. They typically have sound-producing structures, often on the bases of the front wings, and, correspondingly, have ears, situated on the abdomen or on the front legs.

C. Suborder Mantodea—praying mantids. The only exclusively carnivorous suborder of Orthoptera, with the front legs modified into powerful raptorial devices used to capture other insects. Often mantids resemble leaves or flowers. They are mostly tropical, but there are a few striking species in temperate regions.

D. Suborder Phasmodea—stick and leaf insects. These herbivores usually bear a remarkable resemblance to twigs or foliage; they are the largest group of insects to almost uniformly rely on this means of protection. Like the mantids, they are mostly tropical.

E. Suborder Grylloblattodea—ice crickets. These rare and little-known insects are of theoretical interest because they are evidently primitive, combining the characteristics of the other suborders of Orthoptera. Being wingless and furnished with a pair of long terminal filaments (cerci), they resemble some of the apterygotes. The few species have been found in the high mountains of Japan and in the Sierra Nevada and the northern Rockies, most often in ice caves or under rocks at the edge of snow fields.

6. Order Dermaptera—earwigs. Small to large; front wings short, serving as wing covers only, the wide, semicircular hind wings folded both longitudinally and transversely and concealed under the wing covers when in repose; the cerci usually modified into a pair of forceps. This minor offshoot of the Orthoptera, with the specializations noted above, is mainly tropical, although there are some well-known north-temperate species.

7. Order Embioptera. Small; both pairs of wings membranous, narrow, similar in size, the females wingless. These rather termite-like insects live in colonies, in a communal silken web, usually in concealed situations. The group is poorly known, its species seldom seen; most of the species have been described only recently.

8. Order Isoptera—termites. Small; wings membranous, narrow, both pairs usually similar in size, and with few veins, the wings always shed after the dispersal flight. Although they are, like the Orthoptera and other orders so far listed, structurally rather primitive, they have evolved complex societies, and all the many species are social. They abound in the tropics.

9. Order Zoraptera. Very small; both pairs of wings membranous, narrow, hind pair somewhat smaller, the wings usually shed when the insect becomes sexually mature. This is another group of termite-like, colonial species. They have been found on all major land masses except Australia. The group is the most recently discovered (1913) of the insect orders, and the species are rare and remain little known.

10. Order Corrodentia—psocids, bark lice, book lice. Small or very small; wings membranous, narrow, with few veins, the hind pair much the smaller; many species are wingless. This minor order would be little noticed but for the few that are household pests.

11. Order Anoplura—lice. Small; all the species are wingless, although evidently descended from winged ancestors; body form much flattened, an adaptation for a parasitic existence, in which they cling to the bodies of warm-blooded animals. There are two major groups in the order.

A. Suborder Mallophaga—biting lice. With biting jaws, like their orthopteroid ancestors. They feed on hair, feathers, sloughed-off skin, and dried blood from wounds self-inflicted when the host tries to dislodge the parasite. The many species are nearly all parasites of birds.

B. Suborder Anoplura—sucking lice. With the mouthparts modified into a piercing stylet, used to suck blood. The comparatively few species are mostly parasites of mammals.

12. Order Thysanoptera—thrips. Small or very small; wings membranous, very narrow (strap-like), both pairs alike, and fringed with long hairs. The two important features of this order are the invariably fringed wings, of a type that appears sporadically in very small representatives of other orders, and the peculiar mouthparts, which are intermediate between the chewing and sucking types. Thrips are abundant and destructive plant-eaters, but some of the many species feed on small insects or on mites.

13. Order Hemiptera—true bugs and relatives. Small to large; two pairs of wings, which are locked together in flight. In these orders there is a uniformly extreme departure from the biting mouthparts characteristic of most of the orders so far mentioned: they form a hair-thin piercing stylet, used to suck blood or plant juices. There are two suborders, sometimes ranked as orders.

A. Suborder Heteroptera—the true bugs. Basal half of the front wings thickened to form a protective cover for the hind wings and abdomen. These insects characteristically secrete defensive odorous substances.

B. Suborder Homoptera—cicadas, plant lice, scale insects, and so forth. Front wings of uniform texture throughout, often somewhat thicker than the hind wings. Not protected by odorous secretions.

14. Order Plecoptera—stoneflies. Medium-sized to large; both pairs of wings membranous, the hind pair larger and fan-like; the abdomen with a pair of short cerci. Stoneflies are especially characteristic of clear, cold streams and lakes of high mountains, the North, and high latitudes of the Southern Hemisphere.

15. Order Ephemerida—mayflies. Small to large; wings wide, transparent, net-veined, the front larger than the hind pair, or sometimes the hind pair absent; weak, slender, soft-bodied; the young aquatic. These important components of the diet of fish are models

for many varieties of the fisherman's trout flies. The adults, swept from their home waters by breezes and attracted to lights, sometimes accumulate in drifts on city streets. The short-lived adults do not feed.

16. Order Odonata—dragonflies and damselflies. Large; wings long, narrow, transparent, net-veined, the two pairs more or less equal in size; the body long and slender; the young aquatic. These aerial predators are familiar sights near water and also hunt well out over dry grasslands. There are two major groups in the order.

A. Suborder Anisoptera—the dragonflies proper, with powerful flight and the wings held out airplane-like when at rest.

B. Suborder Zygoptera—the damselflies, with relatively weak, fluttering flight and the wings held vertically when at rest.

17. Order Neuroptera—the net-winged insects. Mostly medium-sized to large insects; two pairs of thin, more or less similar wings, which are sometimes weakly coupled at the base. Adults with chewing mouthparts, larvae with either chewing mouthparts or with suctorial mouthparts of a peculiar type. These insects differ from all those so far mentioned in having complete metamorphosis, in which the wings develop internally and there is a pupal resting stage. All of the orders yet to be listed also have complete metamorphosis. The living neuropterans are remnants of once more extensive groups. The three suborders listed here often are considered to rank as orders, in which case the suborder Plannipennia is given the name Neuroptera.

A. Suborder Megaloptera—alder flies and dobson flies. Hind wings with a remnant of the fan-like basal lobe that characterizes the Orthoptera and Plecoptera. Larvae aquatic, and with chewing mouthparts. This small group of insects much resemble the stoneflies, from which they differ by lacking the primitive cerci of that group.

B. Suborder Planipennia—lacewings, ant lions, and so forth. Larvae with peculiar mandibles, each long, sickle-shaped, with a groove covered by the blade-like maxilla, and used for drinking the body fluids of their insect prey. These neuropterans are today the dominant group of the order, and some species are familiar and abundant.

C. Suborder Raphidiodea—snake flies. Larvae with biting mouthparts. The snake flies are given a remarkable appearance by their long "neck" (the prothorax) and the long, sword-shaped ovipositor. The relatively few and usually uncommon species occur on all continents except Australia.

18. Order Mecoptera—scorpion flies. Small to medium-sized; wings membranous, narrow, the pairs equal in size, coupling device, when present, near base of wings; head usually prolonged ventrally into a beak. Larvae with mouthparts of the chewing type; abdomen with caterpillar-like prolegs. The few species of scorpion flies represent an isolated remnant of a once more important group.

19. Order Coleoptera—beetles. Small to large; front wings cupped, hard, thick, opaque, protecting the delicate hind flight wings; mouthparts with biting mandibles, as in the Orthoptera. This great order contains more described species than any other.

20. Order Strepsiptera. These minute insects are a parasitic offshoot of the Coleoptera. The mature female is a sac-like endoparasite of other insects; the male resembles a beetle that has lost the wing covers. The few species are rarely seen by any but the specialist.

21. Order Trichoptera—caddis flies. Small to large; wings membranous, but thickly covered with hairs, the two pairs more or less alike, coupling apparatus rather well developed, more or less restricted to the bases of the wings. Caddis flies are closely related to the next order, the Lepidoptera, and in fact usually look like smallish, dull-colored moths with very long antennae. The larvae of the caddis flies are essentially water-going caterpillars. The most familiar species construct portable shelters made of sand or bits of vegetation tied together with silk.

22. Order Lepidoptera—moths and butterflies. Small to very large; the wide wings membranous, but shingled with scales that are often brightly colored, the hind pair usually subordinate to the front, the two pairs held together by a basal coupling apparatus; mandibles usually missing in the adult, and the other elements of the mouthparts fashioned into a long drinking tube that is carried coiled under the head when not in use. The larvae, or caterpillars, crop vegetation with heavy chewing mandibles and have several pairs of stubby extra legs (prolegs) on the abdomen. An extensive and familiar group.

23. Order Diptera—true or two-winged flies. Very small to large;

the single pair of wings membranous and rather narrow, what were once the hind pair reduced to a pair of inconspicuous short rods that function as balancers; the mandibles often are absent, and the mouthparts are adapted for drinking liquid foods. The larvae of all are without legs, and some are sedentary maggots. Widespread, abundant, and familiar insects.

24. Order Siphonaptera—fleas. Small; wingless, body form compressed; all ectoparasites of birds and mammals, whose blood they drink. The larvae are legless, like those of the Diptera, and are scavengers. A minor group with relatively few species.

25. Order Hymenoptera—wasps, ants, bees, and so forth. Very small to large; wings membranous, narrow, the hind pair much subordinate and the two pairs held firmly together by a distal coupling apparatus; mandibles of the biting type, the remaining feeding appendages forming a structure used to lap up liquid food; ovipositor hard and sharp, used primitively for laying eggs in plant tissues, but in the advanced and familiar members of the order is a sting. The larvae of most are legless, blind, and maggot-like. An extensive group.

These brief introductory descriptions show that two sets of structures are of most importance in characterizing the orders: the wings, and the mouthparts. The classification of Linnaeus was based almost solely on the wings, and his follower Fabricius constructed a classification that was based upon mouthparts. Such relatively simple schemes failed when confronted by the enormous number of insects that soon became known, and classifications subsequently were based upon combinations of characters. Modern classifications still lean heavily on wings, mouthparts, type of metamorphosis, and structure of the legs, but, as classifications are improved, more aspects of insect structure and biology will be used.

Part two

Form and function

3

The outside

The hardened outside layer of the skin of insects is a skeleton, since it gives mechanical support to the animal and provides firm attachment points for the muscles. In the insects and other arthropods, the structure is termed the exoskeleton because it lies on the outside. The term, however, is somewhat misleading, since in fact the skin, and the hardened layer along with it, is often deeply folded into the body to produce internal ridges and pillars that brace the body for added strength and provide more surface for muscle attachment. It is incorrect to say that all of the skeleton of the insect is external.

The hardened layer of the arthropod skin is specifically designated as the cuticle. The entire skin then consists of this structure together with the underlying sheet of living cells that manufacture the dead and inert cuticle (Fig. 13).

The outermost layer of the cuticle is a very thin sheet of wax, and this has the supremely important function of waterproofing the skin —usually not to keep water out but to keep it in, to prevent water loss by evaporation. A simple laboratory experiment demonstrates the function of the wax layer. If houseflies are shaken in a container with finely powdered charcoal, the black dust clings tightly to their bodies. Charcoal has a strong affinity for the wax and absorbs it from the cuticle in such a way as to break up the waterproof layer. Flies dusted with the charcoal—ordinarily a non-reactive, harmless substance—will die in a few minutes when exposed to the air of the laboratory. If, however, flies treated in the same way are kept in humid air, where water can not be lost by evaporation, they will live indefinitely. In nature the layer is liable to abrasion as the insect moves about, but a small amount of such damage can be repaired by fresh wax manufactured by the skin cells and poured to the outside through minute pores.

Underneath the thin waterproof layer, the remainder of the cuticle is a relatively thick layer of protein mixed with chitin (a nitrogen-containing polysaccharide that is characteristic of arthropod cuticle). The protein is usually hard, like the protein of fingernails, except at the joints that allow for movement, where the protein and chitin layer is soft and flexible.

In most insects the cuticle is so rigid that it must be discarded at intervals to allow the animal to grow. This periodic shedding of the cuticle is called molting; most insects molt from two or three to several times before reaching maturity.

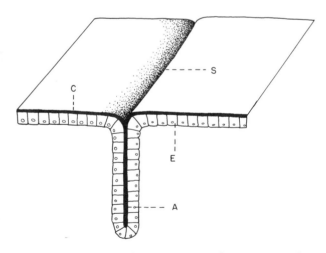

FIG. 13. *Diagram of the skin, or integument, of an insect. A, inflected portion of the exoskeleton, which serves to strengthen the body wall or to give attachment surface for muscles; C, exoskeleton, or cuticle; E, the epidermis, a layer of living cells that produce the cuticle; S, suture, a groove that marks the position of A.*

The first step in molting is to produce a new cuticle under the old one. This new cuticle is soft, white, and flexible and can be stretched. Through pores in the new cuticle are poured enzymes that dissolve away most of the old cuticle on the outside. By swallowing air or by coordinated muscular movements that squeeze blood into one part of the body, the insect bursts the old cuticle, usually through definite cleavage lines on the back (Fig. 14). Since the cuticle extends into the front and hind parts of the digestive tract and even into the tracheae, old skin from here has to be shed also, and the discarded skin is a remarkably complicated piece of castoff apparel, preserving in minute detail the external and much of the internal structure of the animal.

After it crawls out of the old cuticle, the insect is a soft, flabby creature. Almost any large mealworm (*Tenebrio*) culture of the kind sold in pet shops for live animal food will contain a number of ghostly white, newly molted beetles. The freshly emerged insect usually stretches out the cuticle to full size before it hardens. The hardening of the new cuticle is brought about by the diffusing into it of enzymes that are produced by the living cells of the body. These enzymes apparently cause the long, thread-like protein molecules of the cuticle to link up side by side, creating a rigid network of proteins. This lateral bonding, or "tanning," reaction is accompanied by the formation of dark melanin pigments, so that, as the insect hardens, it also darkens. The process takes a few hours.

The insect is quite vulnerable at the time of molting. If the air is too dry, there may be so much water loss that the molt can not be completed properly. The crumpled, aborted wings of the moths that a person has tried to rear in a cage represent such a failure. Insects usually time the molt so that it takes place when humidity is normally high, as in the early morning.

Just after molting, the insect also is vulnerable to predators, being unable to move and lacking armor protection. All in all, molting, particularly the emergence of the winged adult, is one of the most critical periods in the life of the insect, and the biology of this event as it occurs in nature must be much more complex than is now realized.

In the cuticle are the pigments that give insects their brilliant colors.

FIG. 14. *At the close of the underground phase of its life, the cicada nymph crawls above ground and becomes the winged, aerial adult, which leaves behind the empty nymphal cuticle.*

Some colors, especially iridescent blues and greens that change with the angle of light, are caused, not by pigments, but by the microscopic structure of the cuticle itself. Exceedingly fine parallel ridges on the surface will break up white light into rainbow colors by diffraction, or very thin transparent plates will give the same effect by interference, as in the oil film on water.

Also on the cuticle are innumerable minute sense organs, mostly of touch and smell, which will be described in Chapter 7.

Although the cuticle is itself lifeless, its complex form, shaped into the wings and the varied appendages that perform the functions of catching food, of eating it, of digging, walking, or swimming, tells the observer much about the life of the insect.

Of the pairs of appendages that the insect has retained from its many-legged ancestry, the least changed are the legs—three pairs on the thorax—typically used for walking but sometimes for other purposes. Probably the only important change in these appendages since the early, marine, pre-trilobite stage in arthropod evolution is the loss of the feathery gills from the base, or first joint, of the leg.

The leg of an insect is somewhat analogous to that of a human being in that it consists of two long, rigid parts—the femur, which corresponds to the human thigh, and the tibia, which corresponds to the part between the ankle and the knee. In the insect the equivalent of the ankle and foot is called the tarsus. The tarsus usually has five segments, more or less movable on joints between them, and ends in a pair of claws. Usually there is a sticky pad or bristle between the claws, which sometimes are used in hanging on to things. There are two short segments between the body and the femur that have no equivalent in the human leg: the coxa, and the trochanter. These extra segments make it possible for the leg to move in more planes than otherwise, for the fact that the skeleton of the legs is on the outside makes the all-way ball and socket joint like that at the upper end of the human upper arm or thigh impractical.

Some of the leg muscles are in the thorax, others are contained in the leg segments themselves, except that none occur in the tarsus. A slender tendon running from muscles in the tibia through the five tarsal segments moves the tip of the leg.

It has long been known that insects walk in such a way that they put all six legs to good use. As the insect strides along, three legs are left touching the ground—the front and back on one side, the middle on the other—forming a solid tripod, and the other three are swung forward to form another tripod for the next step. Detailed study shows that this ideal may not be quite achieved, since a fourth leg may be left momentarily touching the ground, but essentially walking and running consist of moving from one tripod support to another. If a pair of legs is removed experimentally, the insect readjusts the leg movements so that walking is still possible, although very slow. Only one leg can be lifted and moved forward at a time, this because a tripod support must always be present if the insect is not to fall over. With only three legs walking is not possible. A four-legged mammal, with better balancing mechanisms on account of its greater size, can do better, as the neighborhood dog that whimsically runs with one hind leg off the ground. It is likely that very small animals such as insects can not evolve the balancing mechanisms required for walking or running without tripod support, which requires at least six legs.

When modified for swimming, the legs usually have their movements more or less restricted to a single plane. Also, they present a broad, oar-like surface to the water on the power stroke. This may be done by having the leg flattened, so that it can be turned edge-on for the forward or the return stroke and flat-on for the power stroke. Or the leg may be furnished with combs of closely spaced hairs, so fitted that they hang limply on the return stroke but are forced out into an extended, locked position on the power stroke, presenting a flattened surface to the water.

Digging or burrowing insects usually have the front legs adapted for this function. In the mole cricket (*Gryllotalpa*), for example, the front legs are short and wide, with the thickened femur housing powerful muscles and with the short tibia bearing large shearing blades. The tarsal segments themselves have been altered into similar blades.

The front legs of such predaceous insects as the praying mantids are much modified for seizing prey, sometimes to the extent that they are used little or not at all for walking; on the remaining four legs the mantid can walk only haltingly. Here the femur contains the strong muscles that pull the tibia toward it, and the effectiveness of this trap may be enhanced by sharp spines on the opposing surfaces of both femur and tibia or by felt pads on these surfaces that help hold on to slippery prey.

The thorax itself is composed of three segments, corresponding to the three pairs of legs. The foremost segment, the prothorax, carries a pair of legs only; the following two, the meso- and metathorax, each carry also a pair of wings in most insects. These two segments together form a functional unit called the pterothorax. The pterothorax has large side walls (the pleura) that furnish attachment surfaces for some of the leg and wing muscles and also give the pterothorax the volume necessary to house the large flight muscles. These muscles, and the wings themselves, will be described in Chapter 4.

The head of the insect carries important sense organs—antennae, compound eyes, and ocelli—which will be discussed in Chapter 7.

In addition, the head carries three pairs of appendages that have been modified into mouthparts, used for feeding. Here again, a discussion of the diverse kinds of mouthparts, each adapted in characteristic ways in the different groups of insects for different modes of feeding, will be deferred until the insect groups are taken up individually, but a description of a generalized, rather undifferentiated type of mouthparts will be given here as background information.

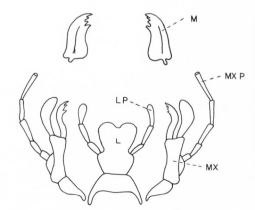

FIG. 15. *Mouthparts of a grasshopper (diagrammatic). L, labium; LP, palpus of labium; M, mandible; MX, maxilla; MX P, palpus of maxilla.*

A set of mouthparts that may be taken as a basic type—like those of the early insects, which can be altered in evolution to produce other types—are those of the Orthoptera, of which the grasshopper is an example (Fig. 15). Such mouthparts, adapted for eating solid rather than liquid food, are called chewing mouthparts.

The design of the chewing type of mouthparts is centered around the mandibles, or jaws, which are the organs that shear off and diminute pieces of food. The mandibles are the most highly modified of the three pairs of appendages, bearing little resemblance to legs. A mandible corresponds only to the basal segment, or coxa, of a leg; in some Crustacea, it is more leg-like, since some of the segments beyond the coxa are retained to form a sense organ called the palpus.

The muscles that move the nearly solid, heavy mandible of the grasshopper are in the head. A very large muscle closes the mandible, a weaker one opens it (Fig. 16). The mandible articulates firmly on two points of attachment, fore and aft. on the head, and these restrict it to a single plane of movement more or less parallel to the face of the insect.

On the inner surface of the mandible, which meets the corresponding surface of its mate, there may be a set of grinding or cutting teeth. Thus, to draw a broad analogy in terms of function, one mandible corresponds to the lower jaw of a human being, the other to the upper jaw, even though the insect jaws work from side to side rather than up and down.

FIG. 16. *Insect mandible, or jaw, and the main muscles that move it. Contraction of the larger muscle on the left closes the jaw, that of the smaller muscle opens it.*

Behind the mandibles are the comparatively delicate maxillae. These are more leg-like than the mandibles: parts of the maxillae called palpi look something like miniature legs and probably correspond to representatives of leg segments that have been lost in the mandibles. The palpi have organs of taste that help evaluate the food. The flattened, leaf-like part of the maxillae near the mouth is used to pick up pieces of food that are snipped off by the mandibles and to pass them on into the mouth. Other parts serve to keep the food particles from falling out at the sides.

Lying behind the maxillae, the labium forms a flap that closes off the back of the cavity in which the food is handled. In essence, the labium is a pair of maxillae fused together in the midline and is therefore furnished with a pair of sensory palpi.

The front of the food handling cavity is closed off by a flap, the labrum, that is fastened to the front of the head; this structure was apparently not derived from a pair of leg-like appendages.

The bulk of the muscles that operate the mouthparts lie inside the head. Especially in insects, such as the grasshopper, having powerful jaws (mandibles), the muscles that close the mandibles may be so large as to fill most of the cavity of the head.

When still an embryo, the insect has a short pair of appendages on each segment of the abdomen, betraying its myriapod ancestry. Some of the primitive wingless insects have several pairs of such structures on the abdomen even as adults, but the rest have discarded all but the appendages near the tip of the abdomen.

Such insect orders as the Thysanura, Orthoptera, Ephemeroptera, and Plecoptera have retained at the end of the abdomen a pair of appendages, called cerci, that function as organs sensitive to touch. These are more antenna- than leg-like in appearance, sometimes being very long and slender and having many segments. The thysanuran (Fig. 34) has not only a pair of very long cerci but also a prolongation of the abdomen that forms a third one. The pair is held out at the sides, the middle one straight backward, so that these three, together with the pair of long antennae on the head, form a warning system, sensitive to touch, all around the insect.

Many of the mayflies (Ephemerida) also have three long cerci

(Fig. 50). In the young mayfly they function as sense organs; in the adult, they trail out behind in flight and may have an aerodynamic function, like the tail of a kite.

The cerci of some of the Diplura and some earwigs (Dermaptera) have been altered to form a pair of pincers used to capture prey and in defense (Fig. 37).

The external genitalia at the end of the male abdomen may or may not be derived from legs—the question is controversial. They are quite complicated structures that are used to clasp the tip of the abdomen of the female during copulation. Their structure varies from species to species, and it is believed that these differences sometimes help prevent individuals of one species from mating with individuals of another species. At any rate, they are often useful to the classifier who is trying to tell one species from another.

The corresponding structures in the female often are adapted for placing the eggs in protected situations. In such orthopterans as crickets and long-horned grasshoppers the ovipositor may be very long and may be used for inserting the eggs in the ground or in plant tissues. In the bees and wasps the ovipositor becomes transformed into a sting. Whether one is bitten or stung depends on which end of the insect is involved in the operation.

4

Flight

The fossil record shows that there were flying insects about 250 million years ago, so that, as far as we know, the insects were the first animals to fly. The first bird appears some 100 million years later, about the same time as the earliest flying reptiles or pterosaurs. Bats, the only flying mammals, appear in rocks of the Eocene age, about 50 million years ago.

It is not known by what stages the insects evolved the ability to fly. The oldest known fossil insect is from the Devonian, and it is a wingless collembollan. There is then a gap of well over 50 million years until insects again appear, and, when they do, in late Carboniferous (Pennsylvanian) times, they are fully winged and apparently with good flight ability. About fifteen hundred species of insects have been described from the late Carboniferous, most of them cockroaches, but including remarkable giant primitive dragonflies with a wing span of up to twenty-nine inches, nearly three times that of any living insect.

Although there are other possibilities, it seems reasonable to think that insects learned to glide before they learned to fly. A gliding insect would have had to be a fairly large and heavy one to maintain the velocity needed for directed glides. Also, it would have had to be able to attain a high initial velocity. This could have been done by jumping from high places, and it has been suggested that flattened cockroach-like insects that jumped from the sides of tree trunks to plane away to a safe landing were the first gliders. It is not easy to see the adaptive advantage of this, since even a large insect does not gain enough velocity in falling to injure itself, but perhaps this behavior would have been useful in putting a good distance between it and a predator.

It is also possible that the first gliding insects were strong jumpers, like the modern grasshoppers. Here any flattening of the body or rudimentary planes at the sides would have had selective advantage in prolonging the leap. From such beginnings there might have evolved extensive fixed sail planes that carried the insect respectable distances. Perhaps in the Devonian, when the first terrestrial vertebrates slithered over the ground, they frightened up "grasshoppers" that sailed away on fixed wings like children's paper darts (Fig. 17).

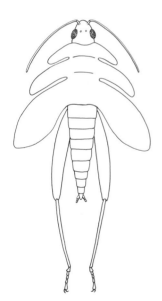

FIG. 17. *Hypothetical first gliding insect. The hind legs are imagined to be powerful jumping legs like those of modern grasshoppers.*

What we do know is that the oldest fossil insects that had any kind of aerial ability, those from the Carboniferous rocks, had two pairs of large movable wings capable of true flight. On the first segment of the thorax was a pair of short, flat, immovable lobes, which may or may not have been relics of an earlier gliding stage. It may be that these lobes provided a small planing surface that helped to hold up the front end of the body when the insect was in flight.

In the evolution of true flight from the gliding stage, it was necessary that the wings become movable. It is possible that gliding insects evolved methods of flexing the long middle and hind planes back over the abdomen to get them out of the way as the insect crawled through tangles of vegetation. With the wings thus movable, perhaps rapid periodic contraction of some of the leg muscles could have pulled on the wing base in such a way as to vibrate the flight planes, producing an aimless, buzzing flight, which would at least have had the merit of carrying the animal farther from danger. This would have provided the basis for the evolution of the mechanically elegant flight mechanisms of living insects.

The Carboniferous winged insects had two pairs of wings that operated independently in flight, and this is true of some groups of insects alive today, but they are in a minority. Most modern insects have in one way or another reduced the number of flight planes to a single pair. In the beetles this was done by changing the front wings into hard, cupped, protective shields that have little or no function in flight (Fig. 56). In the bees and wasps the smaller hind pair of wings are hooked to the front pair so that the two move as a single pair in flight. The Diptera or two-winged flies have converted the hind pair of wings into a pair of diminutive balancing organs (Fig. 65). It is the mechanics of flight in these advanced, often swift-flying insects with essentially a single pair of wings that will be discussed here.

The wing is a hollow outgrowth of the body wall. It is so thin that it appears to be a single sheet of dead cuticle, but, if the wing of a living insect is examined under the microscope, blood cells can usually be seen moving through the narrow space between the upper and lower layers of cuticle, especially near the veins. That blood circulates through the wing can also be shown by applying a drop of insecticide to the wing: the insect may be killed by the toxin that is carried to the body in the blood stream. Also, some of the bristles on the wing are sense organs and are, of course, furnished with living nerve fibers.

The wing membrane is strengthened by thick-walled branching tubes called wing veins. Such insects as the dragonflies, with a large wing area, have a complex network of hundreds of veins, but usually the wings are strengthened by a relatively simple branching system of veins (Fig. 18). The pattern is more or less faithfully repeated in every individual of a species and is highly characteristic of groups of related species, so that above all other structural features it is useful in classification. Probably the different patterns of venation are related in part to differing modes of flight. Relatively slight differences in pattern would affect the way the wing membrane twists during

FIG. 18. *This much-enlarged portion of a dragonfly wing shows the three-dimensional arrangement of the minute struts (veins) and wing membrane that brace the large, thin wings of these expert fliers. Differences in the pattern of the veins in the region of the triangle above and to the right of center are diagnostic of different groups of dragonflies.*

the very rapid strokes of the wing, thus influencing its aerodynamic qualities.

In the more capable flyers the heavier and stronger veins lie close to the front margin of the wing, leaving the posterior half of the wing less heavily braced and more flexible.

The insect wing differs from the airplane wing in that it provides both thrust and lift. In its operation, the narrow, blade-like insect wing functions more like a propeller than an airplane wing. A rotary propeller drives a blast of air behind it or, to put it another way, creates a high pressure area behind and a low pressure area in front of the blades, the difference providing the thrust of the propeller. The angle of attack of the rotary propeller blade is changed by each revolution of the propeller: the leading edge is down on the downstroke, up on the upstroke. Insects achieve the propeller effect by altering the angle of attack of the wing with each phase of the wing beat, so that the insect wing is in effect a vibratory propeller. On the downstroke the leading edge is down; then at the bottom of the stroke the wing is turned on its long axis so that the leading edge is uppermost for the return stroke (Fig. 20). Thus, a high pressure area is maintained behind the wings during both the down- and upstroke of the wings, and the upstroke is not merely one of recovery.

Were the wings to beat in this fashion in a vertical plane, only thrust would be produced, and there would be no component to provide lift. In some insects the abdomen is flattened in such a way that it functions as a fixed airfoil, and the outstretched wing covers of the beetles also may provide some lift, but most of the lift results from the fact that the wings beat in a plane that is inclined rather than vertical. At the top of the stroke, the wing tip is above and behind the base. During the downstroke, it moves down and forward; during the upstroke, up and back. Actually, the movement is not strictly confined to a plane, so that the tip executes a slender figure 8 during each wing beat, with the bottom of the 8 below the head, the top above the abdomen.

That the wing movement drives a draft of air below and back of the insect has been demonstrated by means of an insect in fixed flight. A housefly, for example, that is fastened to the head of a pin with a

FIG. 19. *The corrugations of the outer half of the wing of this scolioid wasp serve to strengthen the wing membrane. In many wasps the wing veins (here confined to the basal half of the wing) extend nearly to the wing tip, in which case there usually are no corrugations.*

drop of melted wax will, when its feet are without support, beat its wings as if in flight. The direction of the resulting air current can be made out by placing the fly near light streamers such as free-hanging threads or, more simply, by holding the fly close to the face so that the draft of air may be felt on the lips.

Both the direction of the wing stroke and the twisting of the wings may be seen by using an ordinary stroboscope, again with the insect in fixed flight. The housefly frequently changes the frequency of its wing beat, so that the stroboscope must be continually adjusted to keep the wings in view.

When the insect is in fixed flight, the inclination of the wing stroke away from the vertical is exaggerated; in natural flight the resulting, nearly horizontal wing stroke is used only for hovering. The greater the forward velocity of flight, the more nearly the wing beat approaches the vertical. In the housefly this has been demonstrated by proper stimulation of the air-speed indicators of the fly. These are a pair of small bristles called the aristae, one on each antenna (Fig. 21). When a jet of air is directed against the aristae, the insect in fixed

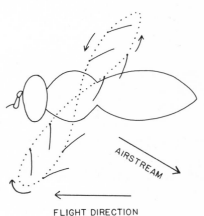

FIG. 20. *Path of a section of the wing during a complete cycle of down- and upstroke in the flight of a two-winged fly.*

AIRSTREAM

FLIGHT DIRECTION

flight responds as if it were moving forward and adjusts the inclination of the wing stroke correspondingly.

By means of the small wing muscles attached near the base of the wing the insect is able to change the angle of attack of the wing and the amplitude of the stroke of individual wings in such a way that it can turn or even fly backward.

The motor that drives the wings is a set of muscles that nearly fill the thorax, so large that they account for a major part of the weight of the entire animal. In the advanced, two-winged fliers that we are concerned with here, the muscles move the wings in an indirect rather than direct fashion. They do so by altering the shape of thorax: the wings are locked into the sides of the thorax in such a way that, when the shape is changed, the wings move. It is a simple matter to demonstrate this with a fly (the common housefly, for example) that is freshly killed or anesthetized. If one places one thumbnail in front of the thorax, another behind, and pushes them together gently, the wing tips move down. This happens because, as the thorax is thus forcibly shortened, the top of the thorax arches upward, and its connection with the wing base is such that the base is pulled up, the wing tip forced down. If one next presses down on the top of the thorax, causing it to straighten and the thorax to elongate, the wing tips move upward.

When the powerful flight muscles inside the thorax contract in the proper sequence, they produce the same changes in the thorax as were produced by the experimenter in the description above. Muscles running fore and aft arch the top of the thorax; others running more or less vertically flatten it when they contract (Fig. 22).

In life the situation is a little more complex but, oddly enough, can be demonstrated with a fly that has been killed with carbon tetrachloride. If one presses down on the top of the thorax of such a fly, the wing tips do not rise smoothly but rather stay motionless until the pressure reaches a certain magnitude and then suddenly flip up. The effect is something like that of turning on an ordinary toggle light-switch: the switch moves a certain distance without effect, then suddenly the contact snaps into place. Apparently the carbon tetra-chloride causes a set of short muscles to contract in such a way that they make the thorax so rigid that the wings cannot move until the force tending to change the shape of the thorax reaches a certain magnitude. This so-called click mechanism of the Diptera operates on both the up- and downstroke of the wings, and results in extremely rapid wing movement.

The fore- and aft-muscles and the vertical muscles are of about the same size, which is as expected if both down and up strokes of the wing deliver power. In this way the flight of insects differs from that of most birds, where the upstroke is relatively weak and the muscles that create the upstroke are small.

The wings of the advanced Diptera and of such Hymenoptera as the bees and wasps beat at extremely high rates (two hundred or more times per second) so that the muscles that drive them must contract at these frequencies, which are unprecedented elsewhere in the animal kingdom. Also without precedent is the structure of these muscles. The microscopic strands, or fibrils, of which all muscles are composed

FIG. 21. *The antennae on the head of this tachinid fly are visible as a pair of bulbous structures on the left. Extending from one can be seen a strong bristle, the arista, that functions as the air-speed indicator in flight.*

are relatively of gigantic size in the flight muscles of these insects, so that a preparation of the pinkish flight muscles from the thorax of a housefly, pressed on a slide under a cover slip, makes a beautiful demonstration under the microscope of the structure of muscle fibrils. Also unique are the mitochondria found in the flight muscles. Ordinarily mitochondria are so small as to be seen with difficulty under the microscope, but those of the flight muscles are so large—

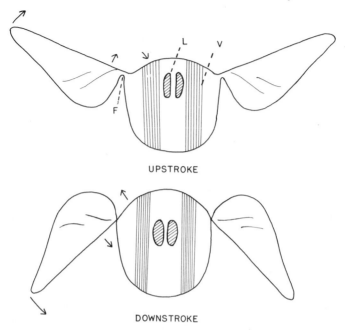

UPSTROKE

DOWNSTROKE

FIG. 22. *Mechanism of wing movement in a typical winged insect. On the upstroke the vertical muscles (V) contract and shorten, depressing the top of the thorax, which moves the wing base down against the pivot (F), forcing the rest of the wing up. On the downstroke the longitudinal muscle (L) shortens, causing the top of the thorax to arch upwards, which pulls the wing base up and forces the wing tip down.*

a third or more of the diameter of a human red blood cell—that they are easily observed. Mitochondria are centers of chemical activity wherein oxygen is incorporated into the biochemical machine in such a way as to yield energy, so that the large and abundant mitochondria are quite appropriate to these remarkably active muscles. When the flight muscles are in full operation, the oxygen consumption increases to as much as fifty times that of the resting animal, and this scale of activity can be maintained for as long as the fuel supply holds out, a matter of hours. A human athlete can increase his oxygen consumption by some twenty times, but only briefly.

The speed of insect flight has been measured by direct observation of insects flying over short measured courses, as in a darkened room with a lighted window at one end. Also, it has been studied by means of a "round-about," a very light rod balanced on a nearly frictionless bearing and carrying the insect glued to one end. Speeds recorded by these methods vary from a few miles per hour for smaller insects (almost five miles an hour for the housefly, six for the honeybee, three for a mosquito) to twenty or twenty-five miles an hour for large hawk moths (Sphingidae) and dragonflies. It has been calculated that a flight of over forty miles an hour would have been possible for *Meganeura*, the giant Carboniferous dragonfly.

It is evident that, with flight speed of only a few miles an hour, smaller insects are at the mercy of even moderate winds. Honeybees, for example, would apparently not be able to forage and then return to the hive if they encountered wind velocities of more than about five to ten miles an hour. Yet insects can be seen flying in controlled fashion, visiting flowers and returning to their nesting sites, even in strong winds. Probably they are able to take advantage of relative lulls near the ground and behind shelters, just as trout are able to move about in streams of seemingly overwhelming swiftness.

To attain even modest velocities insects must move their wings very fast. Wing-beat frequency is measured with a stroboscope, or by tracing the movement of the wing tip on a revolving drum of smoked paper, or by determining the pitch of the humming note made in flight. Wings of the housefly and honeybee vibrate at the rate of two hundred or more times per second, those of larger insects, with relatively greater wing area, beat much more slowly, on the order of thirty times per second.

How far an insect can fly without stopping depends upon the amount of fuel it carries stored in the fat body or in its crop. Butterflies and locusts may fly continuously for hundreds of miles, although, under field conditions, wind velocity and direction more than intrinsic capabilities may determine distance. In the laboratory, the locust *Schistocerca* has flown for more than five hours continuously. The small fly *Drosophila* may fly twelve hours on its food reserves, at a speed of three miles an hour. The fat body of a *Drosophila* that has been flown to exhaustion is without glycogen, indicating that this organ is the site of its fuel reserves. An exhausted fly given a drink of sugar water can resume flight in a matter of minutes. The large crop, a pouch of the digestive tract, of most nectar-drinking insects is a miniature fuel tank that may carry enough carbohydrate for many miles of flight. The honeybee, however, carries only a fifteen-minute supply in its crop, enough to carry it a mile or two, so that the nectar-carrying bee can work profitably only well within this distance from the hive.

5

The inside

As in all active animals, the muscle tissue of the insect makes up a large percentage of the whole body weight. The head is packed with the muscles that operate the mandibles and other parts of the feeding apparatus, and the thorax with those that move the legs and wings. Only the abdomen is relatively free of muscles, and consequently it contains the bulk of the other organ systems: the digestive, circulatory, excretory, and reproductive systems (Fig. 23).

The digestive system converts food into the soluble state needed for nutrients to be absorbed by the living tissues of the body. It is thus a kind of reaction chamber in which the proper admixture of digestive enzymes brings about the dissolution of the foodstuffs.

There are three main divisions of the insect digestive tract, termed, reasonably enough, the foregut, midgut, and hindgut. Both foregut and hindgut are lined with cuticle that is continuous with that on the outside of the body, and, in fact, in the insect embryo, these parts of the tract are formed by simple inpocketings of the skin. It is the midgut that is the dynamic center for the digestive process: here the catalysts that carry out digestion are produced, here they perform their functions, and here most of the soluble nutrients are absorbed.

In those insects which live entirely on liquid food—the cicadas and aphids are examples—part of the foregut has been fashioned into a powerful pump capable of operating over long periods of time

Form and function

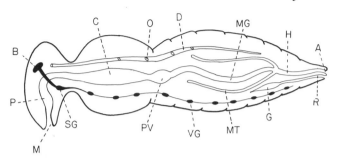

FIG. 23. *Main internal organs of the insect. A, anal opening; B. brain; C, crop; G, gonad; H, hindgut; M, mouth; MG, midgut; MT, Malpighian tubule; O, ostial opening of heart; P, pharnyx; PV, proventriculus; R, opening of reproductive system; SG, subesophageal ganglion; VG, ventral ganglion of nerve cord.*

(Fig. 24). This pump, called a cibarium, is operated by muscles that extend from the wall of the foregut to the front of the head. By their contraction, these muscles enlarge the cavity of the gut, making a partial vacuum into which outside air pressure forces the food. The large swelling visible on the face of the cicada houses the muscles of this pump.

The foregut also is the site of storage for undigested or partly digested food. The storage receptacle is the crop, sometimes little more than a swelling in the nearly tubular foregut, but sometimes a large bulb out on the end of a long tube leading from the foregut and so large when filled that it has to be housed in the abdomen, behind the rest of the foregut. If a starved and thirsty housefly is allowed to drink dyed sugar water, the abdomen quickly becomes colored and distended as the crop fills.

Behind the crop is the proventriculus, or gizzard, which has a thick muscular wall. Sometimes the cuticle lining it is formed into teeth that assist in breaking up the food.

The foregut ends as a short collar-like extension inside the midgut. In the resulting cavity is produced a thin, tubular membrane that extends down the length of the midgut and inside it. This membrane, the peritrophic membrane, thus surrounds the food like a sausage casing. The function of the peritrophic membrane is not completely understood, but its presence might reasonably be related to the thinness of the midgut walls. The fragility of the insect digestive tract impresses the student who dissects these animals for the first time and who compares this delicate structure with the tough, muscular intestine of the frog and other small vertebrate animals. The peritrophic membrane might protect the delicate midgut wall from abrasion by food particles. It also might serve as a tubular conveyor belt to carry the food down the tract. In many insects, a set of teeth in the hindgut engage the membrane and pull it backward, new membrane continually being formed at the front of the midgut. Waves of strong muscular contraction move the food along in the vertebrate intestine, but, because of a basic difference in body plan, a thick layer of muscles is not practicable in the absorptive region of the insect digestive tract. Absorbed nutrients could not go through such a wall; in the

FIG. 24. *Pump used by the cicada in drinking liquid food. MP, muscles of pharyngeal pump; P, pharynx.*

vertebrates, blood vessels penetrate it and bring capillaries close to the inner surface of the digestive tract, where the blood can easily pick up the nutrients. Since insects do not have the blood enclosed in vessels—instead, it bathes the tract and the other internal organs— such a scheme could not be used, and it is necessary that the walls of the insect midgut be thin so that nutrients can get to the blood and to the rest of the body.

As would be expected, the enzymes produced in the midgut are correlated with the diet. Usually there is a battery of enzymes that break down all the three major classes of foodstuffs—carbohydrates, fats, and proteins—but with a restricted diet there may be a more limited set of enzymes. A nectar-drinking adult fly, for example, may be well equipped with carbohydrate-digesting enzymes but is without strong proteolytic enzymes. The enzymes are produced in amounts that make their demonstration in the laboratory a simple matter. The digestive tracts of a few mealworms, ground up and placed in a dilute solution of starch, will in a matter of minutes decompose it to a point where it no longer gives the usual intense starch-iodine staining reaction.

The terminal region of the hindgut is the rectum, where specialized tissues resorb water from the contents of the digestive tract and also recover certain useful salts that are poured into the hindgut by the excretory system.

One of the largest of the internal organs of the vertebrate body is the liver, which carries out the functions of storing certain kinds of nutrients and of converting one kind of nutrient into another as dictated by the needs of the rest of the body. These essential functions are performed in the insect in the cells of the so-called fat body, which sometimes are organized into definite structures, sometimes are scattered more or less loosely in the abdomen. When dissected out alive and placed under the microscope, some of the cells are handsome objects, containing glistening spheres of liquid fat. But the cells of the fat body carry more than fats, some being loaded with glycogen, proteins, and particles of other substances as well.

The blood of insects, containing nutrients absorbed from the digestive tract and the fat body, sloshes about more or less freely in the body cavity, bathing all the internal organs and muscles. It is not confined in arteries, veins, and capillaries. Body movements tend to stir the blood, bringing fresh supplies of nutrients to the tissues, but there is also a pump that creates definite currents through the body, sweeping the blood through the muscles of the head and thorax as well as through the visceral cavity. This pump is the dorsal vessel, a tube that lies just under the back. The thickened, muscular hind part of the tube is the heart, strictly speaking, but the anterior part also is more or less contractile. There are in addition small accessory pumps at the bases of the wings and legs.

Blood enters the heart through paired openings (ostia) in its walls (Fig. 23). It is forcibly drawn in when fibers or muscles extending from the heart to the body wall contract, thus increasing the volume of the heart. To complete the heart beat, the muscles circling it contract. Valves guarding the ostia swing shut, and blood is pressed forward to pour out of the anterior end of the tube, near the brain and through the more anterior ostial openings, sometimes by way of short, open-ended arteries. One can observe the heart in action in the small transparent *Drosophila* larva, where it beats with a rapid flicker. It also is to be observed in careful dissection of anesthetized insects.

The pale blood itself resembles the lymph of vertebrates, since it lacks respiratory pigments and carries only colorless cells. Dissolved in the blood are proteins, amino acids, mineral salts, and sugars, together with other substances. The Malpighian tubules and the resorbing cells of the hindgut, together with the fat body, to a great extent regulate the concentrations of these substances. The make-up of the blood, however, is not as precisely controlled as in the mammals, and the insect will survive drastic experimental alterations of the composition of its blood. The composition varies greatly from species to species and from one stage in the life cycle to another.

The toxic ammonia that is produced when proteins are decomposed during the metabolism of living tissue is converted into uric acid, instead of the urea characteristic of most vertebrate animals. This material is absorbed from the blood by the excretory system, the Malpighian tubules, which are slender tubes, numbering few or many, that extend through the body and connect with the digestive tract at the end of the midgut. Once in the tubules, uric acid is precipitated out as solid particles. The tubules then force the crystals or granules into the digestive tract, where they are finally ejected with the feces. Since uric acid is relatively insoluble, it can be eliminated without much loss of water, a factor important in the economy of insects. Also, it is convenient for storage of wastes by the embryo developing in the egg.

Land animals face a dilemma when it comes to the problem of respiration, the problem of extracting from air the oxygen used in the metabolism of living tissues. To get into the animal, the oxygen must be dissolved in a moist membrane exposed to the air before it can make its way into the underlying tissues. But if a moist membrane is exposed to the air, water will evaporate, putting the animal in danger of death by dehydration. Some insects live in perpetually moist environments, as in the soil, and, if they are small, they can get the needed oxygen by absorption through the moist body surface, which does not lose water in the humid environment. However, most insects do not live in such environments and must solve the problem of respiration in another way.

If the moist respiratory surfaces are folded into the body as pockets, water loss is greatly reduced. The lungs of the mammals are such respiratory pockets; an intricate system of blood vessels then carries oxygen wherever needed. Insects do not have oxygen-carrying blood, nor an elaborate circulatory system, and solve the problem of oxygen transport by having the respiratory pockets in the form of slender branching tubes reaching into all parts of the body. This system of tubes is the tracheal system (Fig. 25).

The tracheal system opens to the outside through several pairs of apertures called spiracles, which are on the sides of the thorax and abdomen. In order to keep out dust particles, the spiracles often are guarded with filters of fine hairs. Also, there are devices that can open and close the spiracles.

Tubes leading from the spiracles to the inside are called tracheae. The lining of these is continuous with the exoskeleton and functions to keep the tubes from collapsing. The lining is in the form of a spiral ribbon: it is possible, with a pair of fine forceps, to grasp the cut end of a large trachea and pull out from this spiral coil an astonishingly long strand of cuticle. Oxygen is not absorbed from the tracheae, but rather from small dead-end tubes called tracheoles, at the tips of the finest branches of the tracheae. Probably no living cell is more than two cell-widths away from a tracheole, and some large cells are even penetrated by them, so that oxygen is readily accessible to all living tissue.

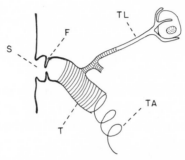

FIG. 25. *Components of the tracheal system. F, filter plate; S, spiracle; T, trachea; TA, spiral thickening of tracheal wall, pulled out; TL, tracheole, which ends (to the right) at a living cell. Components not shown to scale.*

In its simplest form, the tracheal system furnishes oxygen to the tissues simply by allowing diffusion of the gas through the system. As oxygen is used up in the tracheoles, more moves in to replace it by diffusion. It can be calculated that this process will not furnish enough oxygen beyond a distance of more than a few millimeters into the insect, and it is often said that it is the tracheal system, with its reliance on diffusion, that limits the size of insects, that this is why insects are so small.

However, most insects do not have this simple form of the tracheal system and do not rely on diffusion. In a live grasshopper, for example, the abdomen pulsates in a slow rhythm that suggests breathing movements, and in experiments in which the front half and the rear half of the grasshopper are placed in separate chambers, it can be shown that oxygen enters through the more anterior spiracles and leaves by the posterior ones. Evidently there is a ventilating system that drives air forcibly through the tracheae. As the abdomen expands and air is drawn in, the first spiracles are open, the posterior closed; then as air is driven out with contractions of the abdomen, the anterior spiracles are closed, the posterior opened.

In most insects the tracheae are in some places expanded to form large air sacs. When the flight muscles are in use or when the abdomen is expanded or contracted, pressure on these air sacs causes them to act as bellows, driving air through the system.

It is obvious, then, that insects do not rely solely on diffusion for the movement of oxygen in the respiratory system and that other factors must account for their small size. One suggestion that has been made is that it is the necessity for molting that limits their size. While the old exoskeleton is being shed, and for some time thereafter, the animal is a helpless soft blob of material which, above a certain size, would be destroyed by its own weight. It is perhaps significant in this connection that the largest arthropods that have ever lived— the extinct eurypterids—and the largest now living—some of the Crustacea—are aquatic, so that the soft, newly molted animal is buoyed up by the water. The great extinct dragonfly also was presumably aquatic when immature and may have carried out all its molts under-water.

The many kinds of insects that live in fresh water are descendants of air-breathers and usually take up oxygen from the water by means of a modified tracheal system. The various adaptations that enable these animals to meet the special problems encountered in their environment are taken up in more detail in Chapter 12.

The nervous system is all pervading, but, on dissection of an insect, the only part that is conspicuous is the central nervous system, a lengthwise double cord lying on the floor of the body cavity, beaded at intervals with globular swellings called ganglia (Fig. 23).

In the head are two ganglia. The larger, which lies in front of the digestive tract, is the brain, and the smaller, which lies below and behind the tract, is the subesophageal ganglion. These are connected by a pair of nerves that pass on either side of the digestive tract. The brain receives large nerves from the major sense organs: the compound eyes, ocelli, and antennae. The subesophageal ganglion is believed to be made up of three ganglia fused together, because from it arise three pairs of nerves that go to the mandibles, maxillae, and labium.

Perhaps in primitive insects there was, on the ventral nerve cord, a ganglion for each pair of legs on the thorax and for each of the abdominal segments. Most living insects, however, show some fusion or loss of ganglia. In extreme cases, as in the larvae of some Diptera, there is only a single large ganglionic mass, which is situated in the head region.

The ganglia of the ventral nerve cord are rather large, so that it is not surprising to find that they can carry out many coordinating activities independently of the brain. A decapitated insect may be able to carry out complex movements such as walking or those involved in mating. The front leg of a praying mantis, used in seizing prey, will carry out the grasping motion if the segment carrying it and its ganglion are completely isolated from the rest of the body.

Some of the details of the microscopic structure of the nervous system will be discussed in Chapter 7, which deals with behavior.

At the height of the reproductive cycle, the reproductive organs dominate the other internal organs. Indeed, the abdomen of the gravid female may be little more than a receptacle stuffed with eggs, and even the rest of the insect may seem subservient to the function of reproduction: the thorax as a mechanism for transporting the eggs, the head as one for guiding them to their proper destination, a place where the young may grow in safety.

Eggs are produced in the ovaries. Each of the two ovaries consists of a number of parallel tubes, the ovarioles. Immature egg cells are produced near the tip (anterior end) of the ovariole and as they mature are pushed down the tube, increasing in size as the surrounding tissues pour nutrients into them. By the time the egg has received its full store of yolk, it is at the base of the ovariole where the tissues produce the shell. The shell is made of several layers of protein- and wax-containing materials that keep the egg from drying out and at the same time allow oxygen to get in. The shell may be strengthened with ridges or knobs that give it ornate beauty. The shell is put around the egg before fertilization, so that there are small holes, called micropyles, left in the shell at one end of the egg for entry of the sperm.

Two oviducts, one from each ovary, unite to form a tube that, near the tip of the abdomen becomes the vagina. In many insects fertilization of the eggs does not take place for some time after copulation; such insects may then have a special internal organ—the spermatheca—in the female for storing the spermatozoa.

Each of the two testes, which lie inside the abdomen of the male, is made up of one or more lobes or tubes, called the follicles. Two tubes, one from each follicle, unite to form a duct that leads the spermatozoa to the aedeagus, or penis, of the insect. Near the tips of the follicles are sperm cells in the early stages of formation, at the bases are the mature cells. During the development of the sperm their chromosomes become especially easy to observe and count, so that the follicles are much used to study the chromosomes of insects. The number, form, and behavior of the chromosomes are as useful as the structure and functioning of the more easily seen features of the insect body in classifying insects and in understanding their evolution.

6

Reproduction and development

The living individual is ephemeral, is liable to destruction by an infinite number of external and internal causes; a species continues to exist, then, only by the constant production of new individuals.

It is a striking fact that, in the living world, reproduction, or the production of a new individual, is associated with sex—a fact all the more striking when we find that the association is not invariable. In many insects it is possible for a single cell of the body to develop into a new individual. The fact that this cell is an egg produced by a female does not make it true sexual reproduction, which requires that the new individual arise from two cells, each with a different set of hereditary characters and, nearly always, each from a different individual. Many animals simpler than insects can reproduce merely by discarding pieces of the body that grow into new individuals, and, although the much more complex mammals always reproduce sexually in nature, it is possible in the laboratory to start a single mammalian cell—an egg cell—on the road to development of a new individual by stimulating it mechanically or chemically. Reproduction, then, does not necessarily involve sex.

Only since 1900, with the development of genetics, has the biological significance of sex—that sexual reproduction is fundamentally a means of sorting out and recombining different sets of heredities so as to produce variation—been fully comprehended. This outlook

makes it possible to understand why it is in insects that we find reproduction both with and without sex. In some circumstances it is
advantageous for the survival of the species that the population be
highly variable so that there will be individuals available to meet any
of a variety of new conditions. On the other hand, it is sometimes to
the advantage of the species that new individuals be turned out as
quickly as possible, without the time-consuming maneuvers of sexual
reproduction, and that there be little variability.

Reproduction in which there is not the participation of two individuals, where virgin females produce young, is called parthenogenesis.
It may well be that no species of insect is completely parthenogenetic;
and that, even where virgin birth is the rule, there is occasional or
periodic sexual reproduction that serves to mix up the genetic materials
in such a way as to produce variability. Such an insect "has its cake
and eats it too," combining the advantages of quick production of
young with those of promoting variability by sexual union.

The combination of both methods of reproduction is illustrated by
the life cycle of aphids (plant lice), which will be discussed in more
detail in Chapter 15. During most of the growing season only females
exist. These females, without mating, produce female young, which
in turn produce another generation of females, and so on. So rapid
is reproduction here that, even before an aphid is born, it is pregnant,
containing developing young. In the fall, males are for the first time
produced. Females are now fertilized, and each lays eggs, in some
species only one egg per female, which contain the combined genetic
materials of the parents. Thus, in the spring there will be a variable
population of each species, but each local colony, descending from a
single female, that survives into the summer will have little variability
and will more or less faithfully reproduce its type until fall.

Another kind of parthenogenesis is illustrated by species in which
the males are rare, so rare that the majority of females must reproduce
without mating. Such an example is that of the large wasp *Pelecinus
polyturator* (Fig. 10). Males of this species are very rare in North
America and are always a collector's prize. It is interesting that in
Central and South America the males of this species are quite as
common as the females. In those insects where the males are rare,

apparently the advantages of both parthenogenetic and true sexual reproduction are preserved, with the small amount of sexual reproduction that occurs being enough to provide the species with the necessary genetic variability.

Yet another kind of reproduction without sex is characteristic of the thousands of species of Hymenoptera. Here the female, even after she has been mated, has the ability to lay either fertilized or unfertilized eggs. The males always develop from the unfertilized eggs, that is, they are produced parthenogenetically. Females usually develop only from fertilized eggs, although in some Hymenoptera—in the sawflies, for example—both males and females develop from eggs that are not fertilized.

Since insects are primarily land animals, it is not surprising to find that in sexual reproduction fertilization is always internal—the sperm cells are placed in the body of the female, where they are protected from drying. In most insects the male places the sperm cells directly into the body of the female, using the intromittent organ situated at the end of the abdomen. There are, however, variations, as described, for example, in the Thysanura (p. 140) and the dragonflies (p. 186).

Courtship may be fairly complicated, with lures and elaborate displays. Some of the varied courtship patterns will be described as the different groups of insects are taken up. Very often the pattern is characteristic of the species, and a female does not respond to any but that of a male of her own species.

The new individual, produced by sexual reproduction, by the fusion of a sperm and egg cell, begins life inside the shell, protected from drying and able to breathe through minute pores in the shell. This embryo, like that of birds and reptiles, is cut off from aid from its parents. In the yolk are stored enough nutrients to keep it alive and permit growth to where it can break out of the shell and make its own way in the world. The toxic excretory products are, as in the birds and reptiles, converted into the nearly insoluble and therefore harmless uric acid.

The development of the fertilized egg, which at the beginning consists of a single cell, with its nucleus now containing the genetic materials of both parents, begins by the division of the nucleus. With subsequent successive divisions, the nuclei begin to move out to the periphery of the egg cell, and around each the egg cell substance becomes blocked out to form new cells. Eventually a hollow ball of cells is formed, with the yolk in the center. Only the lower side of this ball reproduces the insect. Brain and nerve cord form first, then tissues on each side are blocked out so that the body segments and rudimentary legs, mouthparts, antennae, and eyes quickly become visible and give the embryo the appearance of a miniature insect. It may move about inside the original hollow sphere of cells, leaving it behind as a protective envelope. The digestive tract and the heart form above the ventral nerve cord fairly late in the growth of the embryo, after it has already taken on its insectan appearance.

Although most insects lay eggs and the parent does not aid the embryo as it develops, there are many that give birth to living young, that are viviparous, like the mammals. It may be that the only contribution the parent makes is in sheltering the eggs in the body until they hatch; some of the cockroaches are ovoviviparous in this manner. In others, the developing embryo may absorb nutrients from the mother, thus growing to larger size than if only egg yolk were available. This kind of aid is highly developed in the African earwig *Hemimerus*, where the half-dozen embryos get nourishment through placentae, in the manner of the mammals. Other earwigs, although not live-bearers, care for and brood over the eggs.

If the young insect has chewing jaws, it may simply gnaw its way out of the egg shell. Some are fitted with special equipment whose only use is to help the young out of the egg: a bulging sack of blood that bursts open the shell, or sharp spines that cut through it.

In some insects the young is in appearance a miniature adult. This is especially true of the primitively wingless insects, where there is no difference except in point of size of body and reproductive organs. The young of the more primitive of the winged insects have stubs of wings that increase in size with each molt. Here the young is like the adult except that it cannot fly or reproduce; it may have the same feeding habits and live in the same sorts of places (Fig. 26). The grasshoppers are familiar examples of this kind of development, which is called gradual, or incomplete, metamorphosis. The young are termed nymphs.

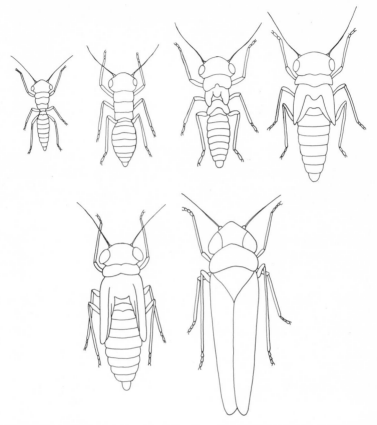

FIG. 26. *Incomplete metamorphosis, shown by the stages in development of a leafhopper. (After Ackerman and Isley.)*

The term gradual metamorphosis is used in contrast with complete metamorphosis, the kind of development that characterizes most species of insects alive today. The bees, wasps, butterflies, moths, two-winged flies, beetles, and some others—varied and advanced groups that made their appearance relatively late in the geological record—have complete metamorphosis, in which the young are termed larvae (Fig. 27).

According to one definition, the word metamorphosis means a change, "as a transformation by magic or witchcraft," a description

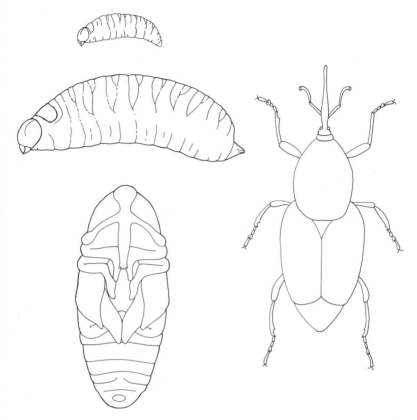

FIG. 27. *Like the butterflies, the Coleoptera undergo complete metamorphosis. The larval stages (the upper left) and the pupa, shown in ventral view (lower left). (After Ayyar.)*

that seems appropriate for the change of the caterpillar into the finely modeled butterfly.

The basic significance of complete metamorphosis is this: that the young is fitted for an entirely different mode of life than the adult. The adult has primary responsibility for reproduction and dispersal, leaving to the young the responsibility for feeding, for converting food into insect tissue as quickly and economically as possible. In no insect with complete metamorphosis does the adult grow; that is, a small butterfly or wasp is not a young one.

The sluggish caterpillar, although much different from the active adult butterfly, is recognizably an insect. It has a well-defined head, with short antennae and three pairs of mouthparts, and a thorax with three pairs of jointed walking legs. But other larvae have none of the external characteristics of insects. The larva of the common housefly and of related flies, for example, does not have a recognizable head, the hooks used to pull food into the mouth have little resemblance to ordinary insect mouthparts, and there are no traces of legs (Fig. 66).

However, if one were to dissect the larva of the housefly, he would find that it is more like the adult than its featureless surface would suggest. Just inside the body wall are rows of dense white knobs that are to become the wings, legs, and mouthparts of the adult fly. These develop inside the body, and, when the larva sheds its skin for the last time, they abruptly turn inside out, pushing out to the exterior.

Complete metamorphosis is characteristic of the four great orders, Hymenoptera, Coleoptera, Diptera, and Lepidoptera, which, together with several smaller orders, include the majority of the living species of insects. The insects possessing complete metamorphosis are sometimes placed in a comprehensive group called the Holometabola or the Endopterygota, the latter term referring to the fact that the wings develop internally.

The stage at which the wings and legs become externally visible is called the pupa. Since the transformation from larva to adult is by no means completed, it is during the pupal stage that much of the metamorphosis takes place. This is, in fact, usually the only function of the pupa. It is usually motionless, does no feeding, and its only activity is physiological and internal, where larval tissues are destroyed and adult tissues built up.

The soft, ghostly white pupa of the housefly is not usually seen, since it is enclosed in a hard, brown shell, the puparium, which is formed by the thickening and hardening of the larval skin. Caterpillars of many moths spin a protective cocoon of silk inside of which they pupate, discarding the thin larval skin in a crumpled heap beside the pupa. Butterflies leave the pupa exposed (this pupa is usually called a chrysalis), and on its surface one can see the outlines of wings, legs, and antennae, all fused closely against the body.

The development of the insect is in part, at least, controlled by hormones secreted from endocrine glands. One facet of this control can be demonstrated with a simple experiment using the housefly larva. If a mature larva is tied tightly about the middle with a loop of hair, only the anterior half will form the hard, brown puparium. If one waits until just before the puparium is normally formed before tying the loop, both halves form the puparium. Evidently some substance diffuses from the front end of the animal that initiates puparium formation: the first experiment indicates that the loop was made in time to cut off the flow of this substance, the second that it was made too late. Further and more complex experiments, involving transplanting tissues from one part of the body to another, show that the source of the substance, a hormone, is a small bit of tissue called the ring gland, which lies near the brain.

7

Sense organs and behavior

In the sense organs living cells are placed in such a way that energy from the environment—the radiant energy of light, the kinetic energy of touch, the chemical energy of reactive molecules that are smelled or tasted—can impinge upon them. The structure of the sense organs is such that only selected influences can act on the sensitive cells. This, together with the fact that the organs are placed on different parts of the body, are connected with a complicated central nervous system and through it with a variety of living body tissues that respond to stimulation by nerves, provide for an unending set of combinations and permutations of reactive elements. The surprising thing is not that behavior patterns of the whole animal are so complex but that, given such a complicated reacting system, the possibilities are narrowed down to the behavioral channels that enable the organism to react meaningfully and successfully to its environment.

The array of sense organs in the insects is a remarkable one. More are still being discovered, and the function of many remains unknown. They are bizarre from the human standpoint because their structure is exceedingly varied and because their location is often unexpected —ears, for example, on the front legs or on the abdomen, and organs of taste on the feet. They may be classified on the basis of the stimulus that reaches their sensory cells: (1) organs of vision, that are stimulated by radiant energy; (2) of touch and hearing, stimulated by

FIG. 28. *Individual elements (ommatidia) of the compound eye can be seen in this close-up of the head of a long-horn beetle. The jointed structures below the head are the palpi of the mouthparts.*

mechanical energy; and (3) of taste and smell, stimulated by chemical energy.

There are different approaches that can be used to get information about the functioning of the sense organs. First, much can be learned from their structure: organs of vision, for example, may be furnished with transparent lenses; of hearing, with drumhead membranes. Second, and more directly, the living animal, behaving in different ways in response to different stimuli, or with sense organs mutilated, gives information on the capabilities of its sense organs. Such studies, based on behavior, have the disadvantage that there are many internal events between stimulation and the response. This difficulty is partly met by a third and extremely recent method that uses sensitive electronic devices to record directly the impulses coursing through the nerve fibers that lead from the sense organs.

In some thin-skinned, more or less transparent larvae there are groups of sensory cells lying under the skin that constitute very primitive light-receiving organs, but more typical are the highly evolved eyes of adult insects, in which a transparent refracting lens covers the light-sensitive cells beneath.

The insect usually has two sets of eyes, a pair of large, laterally placed compound eyes and, between them, three smaller, single-lensed eyes, the ocelli. It is only the first that are "eyes" in the usual sense of the word, since they alone can form and record an image. So far as known, each ocellus can allow for the perception of a single patch of light of presumably uniform intensity.

The compound eye is made up of many separate visual units. On its surface are a large number of small hexagonal lenses, fitted together like the cells of a honeycomb (Fig. 28). Each lens is a transparent bit of cuticle that, together with a cluster of transparent cells beneath it, bends light rays to a focus on the sensitive cells. Beneath each lens is a cluster of sensory cells. At the core of the group of sensory cells, and formed in part from each of the cells, is the actual light-absorbing structure, the rhabdome, a pigmented rod that is oriented perpendicular to the surface of the eye. The lens, the sensory cells, and a sleeve of boundary cells together make up a single visual unit, the ommatidium.

The compound eye is a rigid structure, without the mobility or the capacity to focus of the vertebrate eye. The only movement is one analogous to the widening or closing of the iris, in which the pigment cells of the sleeve around each ommatidium may expand or contract according to intensity of light. In strong light they are expanded, thus isolating each unit from its neighbor. · The light passing through a lens thus falls only on the sensitive element directly beneath it. The resulting points of light, one for each visual unit, together form a mosaic image. In dim light, the sleeves are withdrawn forward, allowing light passing through a single lens to fall on more than one rhabdome. Since little light is then lost by absorption on the black curtains, most of the light falls on the retina, but a relatively poor and diffuse image is the result.

The sharpness of the mosaic image produced on the assemblage of rhabdomes—the retina—of the compound eye depends upon the number of ommatidia. With few visual elements, the image is a coarse one. The dragonflies, that hunt their winged prey in flight, have the largest number of ommatidia, about twenty-five thousand in each eye. The housefly has about four thousand, and there are many insects with only a few hundred.

With their relatively small number of sensitive elements—thousands of rhabdomes instead of the millions of rods and cones in the vertebrate eye—it is not surprising that the visual acuity of insects is poor compared to that of vertebrates. A photograph of the image produced by the compound eye of a firefly gave a resolution estimated to be about 1/75 that of the human eye. The sharpness of vision may also be tested by moving a pattern of stripes of varying widths in front of the insect, noting how wide the stripes must be to provoke a response. The honeybee is thus shown to have about 1/100 the visual acuity of man, *Drosophila* about 1/1,000.

Judging from the selection shown in the flower-visiting habits of honeybees, they can discriminate form, and this can be verified experimentally. Food is presented on a plaque with a given design painted on it, and after a time the bee learns to come to such a plaque, ignoring others, even when no food is present. By presenting trained bees with a variety of patterns, it can be determined, for example, if the bee can distinguish a solid circle from a solid square. It turns out that the bee can not distinguish between the two solid shapes but can distinguish between solid and broken patterns, between a hollow and a solid square, for example.

In like manner, honeybees can be trained to alight on plaques of different colors. Such experiments show that the honeybee is blind to red, can distinguish colors fairly well in the region of the greens and blues, and can see ultraviolet that is well beyond the human range of vision. This range is characteristic of insects in general—most are insensitive to waves longer than 6,500 angstrom units, and many can see ultraviolet light as short as 2,500 angstrom units. The difference between insects and man in the ability to see light of long wave lengths makes it possible to design outdoor lights that will not attract insects. These are yellow (rather than red) so as to give high illumination for human beings but are still near enough to the red end of the spectrum to be relatively invisible to insects.

The near-sighted focus of the compound eye can not be changed, and this, together with its poor definition, would seem to indicate that insects are poorly equipped for vision. However, the eye is apparently very well adapted for the perception of movement. Flickering patterns over the mosaic of the compound eyes, as either the insect or the object moves, perhaps give a dimension of visual experience incomprehensible to us. It is movement that is quickly perceived by insects, a fact learned by the four-year-old girl whose older brother had appropriated the family butterfly net and who with patient approach managed to catch butterflies "by hand!," as she triumphantly announced.

The two large, laterally placed eyes of the insect seem normal enough, but the three additional "eyes," set in a triangle on the forehead, are without analogy in the more familiar animals, although the ancient "pineal eye" of the vertebrates comes to mind here.

The function of these extra eyes, the ocelli, is little understood. Each ocellus consists of a single lens over a light-sensitive retina. The retina is made up of so few sense cells that any image perceived could be only a very coarse-grained one, but the lens focuses well behind the retina anyway, instead of on it, so that the ocelli cannot operate in the way that a lens-equipped eye usually functions. That is, it can not be a sense organ that makes it possible to perceive form. When ocelli are blackened, the insect may cease to respond to stimuli, so that the rather meaningless generalization has been made that stimulation of the ocelli gives "tonus" to the neuromuscular system. This leaves unanswered the questions of why are they absent in some insects or what is the adaptive significance of this particular method of getting the neuromuscular system into working condition.

From their distribution within the winged orders of insects, it is obvious that the ocelli have something to do with flight. When, as often happens, there are both species with and without ocelli in a group of related species: ocelli are present in winged forms, absent in wingless ones. There are some complexities in the situation, however. The Coleoptera generally lack ocelli, although many tens of thousands of species of beetles are good flyers, and many of the strongly flying moths are without ocelli. Perhaps the fact that many of the beetles and moths are nocturnal flyers is relevant. It is interesting that the family Pyrgotidae, in the Diptera, is nocturnal and lacks ocelli, whereas closely related families are day-flyers and have the ocelli characteristic of the order. The reverse situation exists in Hymenoptera, where species that fly in the dim light of dawn or evening, or at night, have ocelli that are unusually large (Fig. 29). Perhaps the ocelli are orienting devices used to maintain stability in flight.

Insects have tactile hairs, bristles, and spines on the general body surface and on the appendages, and these are movable by reason of being set into thin flexible discs of cuticle. When one of these projections is moved by touch, its base presses on one or more sensory cells, thus stimulating them mechanically. The larger bristles or spines are not scattered at random but are precisely situated, and the exact location of these important sense organs varies from group to group, so that they are used in the detailed classification of some insects.

Of most general importance are the small and numerous sensory hairs on the antennae. The antennae are held out in front, as the arms of a man moving through a dark room, and are constantly in motion as they test the surface of objects that are encountered. The more primitive insects have also a rear guard, a pair of antenna-like cerci

FIG. 29. *One of the three simple eyes (ocelli) of a nocturnal scolioid wasp reflects the light of the electronic flash used for photography; the other two are not easily seen in this photo. The ocelli of nocturnal Hymenoptera and of those that fly in the dusk and predawn light are often larger than those of other Hymenoptera.*

fitted with sensory hairs. The cerci, in contact with the earth, detect the footfall of an approaching predator. Anyone who has walked through a field of singing crickets has noted the silence that falls around him.

Other sensory cells that respond to mechanical stimuli are attached to minute domes of very thin cuticle. These structures (campaniform organs) are found on the wings, appendages, and the general body surface. One theory of their function is that they are stimulated when the cuticle bends. Those at the bases of the legs, for example, would be deformed when the leg bears the weight of the insect. When the dome buckles outward, the attached sensory cells are stretched and stimulated. These and a variety of internal sensory cells that also are stimulated by stretching give information about the position of the insect rather than about environmental events and thus function in bodily coordination.

In all insects there is in the second segment (pedicel) of the antenna a cluster of sensory cells called Johnston's organ that extends from its walls to the base of the third segment. The pedicel and the following segments of the antenna (the flagellum) are moved as a unit, and, when the flagellum strikes against an object as the antenna is moved, there is shearing between the pedicel and the third segment. The resulting stress stimulates the cells of Johnston's organ.

The more delicate sensory hairs may respond to sudden air currents, as those, for example, reproduced by the movement of a nearby predator. Sensory bristles on the wing veins and the front margin of the wings may, by their response to air currents, provide information used to control flight movements. Such delicate sensory hairs can function also as hearing devices. Some caterpillars respond by sudden movements to noises, and that the body hairs are the receptive organs is shown by the fact that the insect no longer reacts when the hairs are "deadened" by condensed moisture.

Some insects have a good sense of hearing, always associated with the fact that in them sound plays an important role in mating. These include certain flies, such as mosquitoes, where the antennae function as ears, and the cicadas and some Orthoptera that have highly evolved ears situated in the abdomen or on the front legs. When the antennae are used in hearing, as in male mosquitoes, they are clothed with long hairs that are set into vibration by the whine of the wings of the female. The vibrating antenna stimulates the sensory cells in the pedicel.

The very keen ears of cicadas and Orthoptera—about as sensitive as the human ear—have a taut drumhead membrane exposed to the air. The sensory cells attach to this or to another membrane that resonates with it. Apparently the insect ear cannot distinguish pitch, but it is quite sensitive to differences in intensity and patterns of sound, which makes it possible for the insect to recognize the courtship song of its own species, for example.

The ancient senses of smell and taste are alike in that the sensory cells of both are stimulated by molecules that react with them chemically. A difference between them is that taste is concerned with relatively large amounts of materials and distinguishes only a few qualities, while smell is concerned with minute amounts and distinguishes an apparently endless series of qualities. These sense organs resemble the organs of touch. They include bristles, hairs, cones, and domes, all constructed of very thin cuticle and often sunken in pits that shelter them from mechanical stimulation. The difference between a sense hair that is a chemoreceptor and one that is a mechanoreceptor is that the first is rigid (not mounted in a movable socket) and that extensions of its sense cells extend up into the hollow core of the hair. Presumably the reactive molecules that are smelled or tasted pass through the thin wall of the hair, or through holes in it, and come in contact with the sensory cells inside. There are numerous sense cells associated with each sense organ.

The sense of smell is sometimes exceedingly well developed in insects. A female moth releases a minute amount of a sex attractant into the air. A mile or more downwind a few or, perhaps, even a single molecule of this substance comes in contact with the antenna of the male and is perceived, causing him to turn upwind and eventually track the perfume to its source.

When the insect has a remarkably keen sense of smell, it usually turns out that the antennae are modified in such a way that the surface, studded with microscopic sense organs, has been much enlarged by means of feathery branches or flattened plates.

Because the honeybee can be trained to associate food and scent—that is, to come to a given odor with the expectation of finding food—it reveals to the observer its olfactory abilities and is therefore much used in experiments on the sense of smell in insects. The honeybee has good discrimination and memory in this respect, since it can be trained to seek out a single fragrant odor to the exclusion of forty others. In the hive, where the odor of the flower forms part of the communication system whereby the successful forager informs other bees of a good source of honey, this ability is of obvious importance. Interestingly enough, the odors normally attractive to bees are pleasant to man, and those repellent to bees are repellent to man. It is not possible to train bees, even by rewarding them with food, to come to the evil-smelling compound skatol. Other insects, however, are attracted to such repellent odors, and some flowers that are pollinated by carrion-frequenting flies secrete compounds that smell like the odors of decay.

Organs of taste are found, naturally enough, on or near the mouth, especially on the short, movable, antenna-like palpi. However, the most sensitive taste organs are those on the tarsi. These organs can be demonstrated easily by touching the tarsi of a housefly or butterfly with a brush moistened with sugar water. The fly immediately responds by thrusting out its short proboscis, the butterfly by uncoiling the long drinking tube. One species of butterfly could, when well-starved, taste the sugar in a solution far more diluted than could be tasted by a human being. Apparently the four qualities of bitter, acid, salt, and sweet are discriminated by insects.

The insect continually cleans off the surfaces that carry sense organs, as a man polishes his glasses. The ground-dwelling cockroaches, for example, spend a good deal of time grooming themselves, and this behavior makes them vulnerable to such poisons as sodium fluoride dust, which they get into the mouth while licking their feet clean. Many of the hairs, spines, and brushes on the legs of insects (often used by the taxonomist to classify insects) are used in stylized ways to clean off this or that part of the body. For example, on the tibia of the front leg of bees and wasps is a curved, comb-bearing spine that faces a hollow in the base of the tarsus. The antenna is drawn through this equipment to clean its innumerable sense organs.

The coordinating or nervous system is connected with sense organs, muscles, and glands. The connecting units are the nerve cells, furnished with long fibers that correspond to the wires of an electrical circuit. Nerve cells are of three kinds: (1) sensory cells, whose cell bodies lie in or near the sense organs, and whose long fibers lead to the central nervous system; (2) association nerve cells, whose bodies lie in the central nervous system and whose fibers connect with other nerve cells; and (3) motor nerves, connecting with muscles or glands. Each cell has, besides the long fiber, a short one that receives stimuli from other nerve cells. The actual connection between nerve cells, whose fibers do not quite touch, is a chemical one rather than physical one and acts as a valve, allowing stimuli to go only one way. This is one of the factors producing order rather than chaos in the operation of the nervous system. These chemical valves are vulnerable to certain phosphorous-containing poisons that not only are some of the most potent insecticides but are related to the highly toxic nerve gases of chemical warfare.

In nature the sense organs are continually bombarded with a variety of stimuli, but the insect cannot react to each stimulus as a unit and in a mechanical way if it is to survive. Rather, the stimuli have to be evaluated in terms of such factors as relevance or priority, with the result that many stimuli seem to be ignored. Also, a given stimulus will in different situations produce different responses. The nervous system thus has an integrative function that gives to behavior its "purposeful" or adaptive quality.

The inhibitory mechanism that shuts out irrelevant stimuli may be a chemical one rendering inoperative whole sections of the nervous system. Perhaps this approach affords an objective definition of an emotion or an emotional state as a state in which inhibiting (chemical) mechanisms channel behavior with intensity along a single, restricted line. If emotion is channeled behavior, one can say with certainty that insects have strong emotions: the bee in the hive that has given up its load of field-gathered pollen now has its behavior channeled toward getting back, across a wide expanse of remembered territory, to its patch of flowers; and, once there, it scrambles hastily from flower to flower, quickly gathering pollen and probing for nectar. With its crop and pollen baskets full, the odor and sight of the flowers suddenly means nothing; is the honeybee now filled with an intense longing to get back to the crowded, wax-smelling hive? At any rate, it acts as if it is and heads in a "beeline" for home.

Examples of different inhibitory patterns, and hence different behavior patterns, that are set up under different situations can be cited in other insects. Thus, a species of thrips (Thysanoptera) that lives inside small, tubular flowers normally avoids light, a reaction that keeps it inside the flower, where it feeds; but, when disturbed by mechanical stimuli, as when the flower is shaken, its reactions toward light and the interior of the flower are reversed, so that it flees to the outside, falls to the ground, and escapes. Houseflies normally avoid brightly lit areas, but, if one slaps noisily at them with a rolled newspaper, they fly towards the sunlit windows.

Before it encounters the outside world, each individual goes through a complex internal history in which the fertilized egg begins the sweep of development that will produce an individual able to secure its sustenance from the environment. Development and growth continue after the insect hatches, but added to this is another process that captures our attention, the unfolding of its behavior pattern, its external history, which, aside from minor variations, is, like the embryonic history, essentially the same for every individual of the species.

The description of life histories and the associated patterns of behavior properly belongs with the accounts of the various groups of insects, but some more general aspects of such behavior patterns can be taken up here.

The flexibility of behavior shown by some of the mammals (man, in particular) that depends on experience and learning is based on a large central nervous system. The insects, because of their small size and necessarily small central nervous system with a small number of components, must rely on built-in or instinctive behavior patterns that give it a stereotyped but accurate response to a limited number of environmental stimuli. The behavior pattern is thus largely inherited, just as the structure or physiology of the body is inherited.

This may be another of the many reasons why there are so many kinds of insects. In order to live in a specific habitat, such as the inside of a seed, or in another insect, or in a temporary pond, the insect must have a behavior pattern that enables it to find and successfully live in this habitat. If two insects adapted to different modes of life mate, their offspring, since this instinctive behavior is hereditary, will likely have an intermediate behavior pattern fitted to no available mode of life, an inheritance that would be even more disabling than the intermediate physical features resulting from such a cross. There is, then, great pressure for populations differing in adaptive behavior to evolve mechanisms—usually concerned with mating—that keep them from crossing. And when such a barrier is evolved, a new species is produced.

The behavior of insects is more predictable than that of the vertebrates, a fact obviously related to their smaller nervous system. For example, if one of the compound eyes of a certain species of fly is covered, the individual invariably responds by walking in a circle, with the darkened eye innermost; and other species of insects react in other characteristic and stereotyped ways if an eye is covered. There are also other reactions that give insects the appearance of mindless automatons. A honeybee that accidentally works its way through a narrow opening into a house will endlessly fly towards a lighted window in vain efforts to escape. If a cockroach is handed an antenna from another cockroach, the "machinery" for cleaning antennae is set into motion, and it cleans the foreign member. The moth flies helplessly into the light, even though it is burned to death.

On the other hand, the insect in nature behaves in a more competent manner. The honeybee remembers the terrain in an area a half-mile or more wide, and can find its way unerringly through a wood to its hive. The moth finds nectar-laden flowers in the dusk, drawn by the feeble light they reflect.

These aspects of insect behavior have produced much discussion as to whether or not insects possess "intelligence." It is easy to smile at the honeybee hitting away futilely against the windowpane. Yet some extraterrestrial observer would probably find mass human behavior astonishingly inadaptive and inappropriate, not to say suicidal. It is only that it takes a little more to confuse a human being than it does an insect. The point is that there are degrees of intelligence, dependent primarily on the size of the coordinating system.

Some examples of learning have already been mentioned in connection with sense perception in the honeybee. An example showing both "intelligent" and "nonintelligent" actions is the behavior of ground-nesting social wasps (*Vespula*) when their nesting site is disturbed. These wasps nest in a large underground cavity with a single narrow entrance. The wasps bringing in dead insects to the young find the entrance by using some prominent stone or weed near the nest as a landmark. If the landmark is moved to a position a few feet away, the homing wasps congregate there in great confusion. They are unable at first to cope with the new situation, but their disorientation lasts only for a time, and they finally relearn the landmarks, thus adjusting to the new state of affairs.

Some of the simpler behavior patterns, involving a comparatively fixed response to one environmental factor, can be observed in the laboratory. Larvae of the housefly, for example, always move away from light. If these are placed on a sheet of paper, then doused with a water-soluble dye, their path across the paper is recorded. Shifting the light causes the path of the insect to shift correspondingly. The sensory basis of this response is relatively simple. A group of light-sensitive cells in the "head" of the larva is shielded by a dorsal black plate. Light coming in from the side stimulates the cells, and the insect changes its orientation until the plate shades the cells, thus keeping pointed always away from the light. This response is adaptive, since it keeps the larva from wandering out of the decaying masses of material in which it burrows and feeds.

Part three

Insects and their environment

8

Climate and season

The part of the earth inhabited by insects is a complex mosaic of differing conditions of light, temperature, rainfall or snowfall, and of kinds and amounts of vegetation. This means that a given species of insect will be restricted in its distribution to certain areas of the earth. More than this, the tilt of the earth's axis imposes a yearly cycle of seasons, even in tropical areas where, instead of cold winters and hot summers, there are wet and dry seasons. This means that the insect must adapt to a set of drastically changing conditions, and, since it lives usually for only a single year, the phases of its life are those of the seasons.

Over most of the continental land masses, winter or long seasonal droughts make it impossible for insects to be active the year round. Insects therefore must be able to endure long unfavorable periods. This they accomplish by finding or constructing a shelter and becoming quiescent, so that they expend little energy. The long quiescent period is termed diapause, whether it occurs during cold seasons (hibernation) or during dry seasons (aestivation). It may occur during any stage in the insect's life history, from egg to adult.

Diapause is one of the most complex aspects of insect physiology. It may be so deeply ingrained that, even when the insect is brought into the laboratory and given continuously favorable conditions (warmth, moisture, food), it may nevertheless become inactive at the same time as it would have in nature and may remain dormant for a long period, again becoming active at about the time it would have in nature.

Facts of this sort illustrate two interesting features of diapause. First, the factor that initiates diapause is often not necessarily the one (cold, dryness, lack of food) that makes it desirable for the insect to become dormant. Second, there is evidently a physiological "clockwork" mechanism that secures correlation with the seasonal cycle by keeping the insect dormant for the right length of time. The adaptive value of the first arrangement is usually this: that the insect must prepare in advance to meet the unfavorable conditions. Shelter must be found or constructed before immobilizing cold sets in, or it is advantageous for the animals to stop eating before the food supply becomes seriously depleted, as this might cause them to have to enter diapause in a state of poor nutrition. The adaptive value of the "clockwork" mechanism is that the insect does not have to rely on untrustworthy environmental factors. If an insect egg, for example, were to complete its development and hatch as a result of being warmed by midwinter thaws, the result might be disastrous.

Just how an insect coordinates its life cycle with the seasonal cycle has been shown in studies of the Oriental fruit moth. The caterpillar of this moth lives in the twigs and fruit of peaches and other fruit trees. The moth comes out in early spring and lays eggs which in less than a month go through a complete life cycle. One generation quickly follows another, there being as many as five in a year. But in the fall a caterpillar of the last generation, instead of transforming into a moth, becomes dormant inside its cocoon and remains so until the following spring, when it changes into a pupa, then into the adult moth, to begin another year's cycle.

Experimenters tested the hypothesis that changing diet (ripening fruit in the fall, for example) might bring on diapause, but the caterpillars, which were kept in outdoor cages, became dormant, no matter what their food, at the same time as their wild neighbors. Cages were brought indoors where the temperature could be changed at will, with the idea that the characteristic fall temperatures when imitated in the laboratory might induce diapause, again with negative results.

Success in forcing the caterpillars to become dormant at the will of the experimenter came only when the caterpillars were grown under artificial light where periods of light and dark could be controlled. If caterpillars were exposed to light for eleven or twelve hours out of each twenty-four, all became dormant after spinning the cocoon and did not transform into moths until a few months later. With either more or less light the caterpillars did not become dormant. The eleven- or twelve-hour light period corresponds to the day length of the time of year in which the caterpillars in the wild go into diapause. In practice, it is usually found that temperature has some influence on the effect of light on diapause, but light is the dominant factor.

Unlike other seasonal variables, length of day (photoperiod) is absolutely reliable as a calendar and would thus seem to offer the most desirable cue for adjustment to the seasonal cycle. It is interesting to see that very many species of both plants and animals have been shown to be sensitive to photoperiod, which influences not only diapause but also many other aspects of behavior.

Not only must the insect become dormant at the proper time, but it must also break diapause and become active at the proper time. The overwintering egg must hatch when young growth of the right kind of plant is available for the young caterpillar: the dormant parasitic wasp must be roused to activity at the time its host insect is available; the bee must dig out of its underground winter shelter when the flowers of the proper species of plant are yielding nectar and pollen. Usually the physiological clockwork that terminates diapause operates at low temperatures. The nature of the low-temperature reaction or reactions is not known, but it is probable that the slow decomposition or synthesis of some substance proceeds until a threshold concentration that ends dormancy is achieved.

Most insects have a life cycle that corresponds to a single seasonal cycle; they live but a year. Many, however, are specialists in rapid growth and reproduction and may have several generations during even a short, northern growing season, before cold weather makes it necessary for the last generation of the summer to become dormant. Other insects—as, for example, some that live in houses—are able to be active throughout the year, and generation may follow generation at the rate of several a year without pause.

Some insects require two or more years to reach maturity. Typically, these live in water, under bark or in wood, or in the ground. Examples are: certain mayflies, whose aquatic nymphs may need three years to become full grown; various wood-boring beetles (Cerambycidae, Buprestidae), that may require several, perhaps ten or more years to mature—and there are records of such beetle larvae believed to be forty to fifty years old; and the periodical cicada, whose subterranean nymph slowly grows for seventeen years before emerging and transforming into the adult that will perish within a few weeks. These long-lived insects exist in stable environments that are relatively isolated from the day-to-day changes that impinge on most insects; their pre-adult life is a long and, one would suppose, a dull one.

The soil shelters countless insects during their winter dormancy. Here they are protected from predators, from drying, and from extreme low temperatures. In one locality in Minnesota, while the mean daily air temperature during the middle part of February varied from $4°$ to $-17°$C., the soil temperature at a depth of two inches varied from a fraction of a degree below zero to $-7°$C. and at a depth of twenty-four inches ranged from just under zero to $-2°$C. Both sod covering and a blanket of snow further ameliorate temperature fluctuations in the soil.

Temperature and moisture conditions under fallen logs or stones, or deep in rotten wood, are much like those in the soil, and here the collector in winter may find a great variety of insects. Other shelters are hollow stems of dead herbaceous plants, or plant galls. Many insects construct shelters, such as cocoons. Perhaps insects inside small shelters of the last kind may tend to smooth out fluctuations of temperature by reason of the high specific heat of their body fluids. The cocoon also protects its occupant from drying winter winds.

By seeking shelter in the soil or in other protected places, the insect avoids extremes of cold, but it may not succeed in evading temperatures that drop below $0°$C. However, it has physiological devices that enable it to survive such low temperatures. One can find insects overwintering under a log that are encrusted with ice crystals, and, although at first immobile, they will revive and move slowly when warmed in the palm of the hand.

Although the freezing point of water is 0°C., if it is cooled slowly and without mechanical disturbance, it can be chilled (supercooled) to far below zero without freezing. When freezing does start—often from a single small center—ice crystals spread quickly through this supercooled liquid and its temperature rises to zero. Many insects survive the cold with their body fluid in this supercooled condition in which there will be no ice formed inside the body. For reasons not understood, supercooling is more easily maintained if the insect blood contains glycerol, a compound which has been used in hospitals to preserve frozen blood. The body fluids of insects in hibernation contain this substance.

Some cold-hardy insects can withstand freezing, the formation of ice crystals within the body, and do so with the aid of glycerol. Such insects are not completely frozen, however, since very low temperatures—below $-40°$C.—cause additional ice formation that kills the animal.

The resistance to low temperatures shown by insects in the laboratory is about what would be expected from their mode of life. The flour beetle, *Tribolium confusum*, is a pest of grain products in flour mills or kitchens and normally is not exposed to very low temperatures. If chilled to 7°C., the beetle becomes immobile and dies in a few weeks, even though no freezing takes place. Insects that live in the water rarely are exposed to temperatures that drop much below 0°C., even if frozen in ice, and laboratory studies show that these do not tolerate temperatures much below zero. It is the insects that overwinter in exposed situations that often show extreme cold-hardiness. For example, the pupa of the large promethea silk moth, in a cocoon that hangs on a twig through the winter, will survive long exposure to $-35°$C.

The speed and economy of flight makes it possible for winged animals to escape the cold by moving to lower latitudes. Since migration by small animals would likely be unnoticed, the use by insects of this kind of adjustment to the seasonal cycle is probably underestimated. But there are many known examples, particularly among the butterflies, that regularly migrate with the seasons. The familiar monarch butterfly (*Anosia plexippus*) in North America lives during the summer in regions where it is unable to endure the winter, and those that emerge from the pupae during the late summer months fly southward. In late August of one exceptionally favorable year for monarchs, a person driving along the highway that skirts the south shore of Michigan's Upper Peninsula could see every few seconds for mile after mile a monarch beating its way southward, out across the narrow tip of Lake Michigan. The migrants overwinter in the southern United States but do not return the following spring. It is their descendants, reared in the southern regions, that fly northward, and by full summer the gaudy, striped monarch caterpillars are feeding on the milkweed of the North.

Very few species of insects are of world-wide distribution. Most occupy a restricted territory, usually well within a single continent, sometimes within a few thousand square miles or, very rarely, within even a few square miles. The factors that prevent their indefinite spread are usually ecological, although such grossly mechanical barriers as the oceans may limit them. The number of factors that are involved is indefinitely large: unfavorable temperatures, too much or too little rainfall, unfavorable soil conditions, absence of a necessary host plant, presence of predators or of competitors—any of these and more can make the surrounding regions uninhabitable.

It is obvious that, if the range of a species is an extensive one, extending over several degrees of latitude and longitude, different limiting factors would hem it in on different sectors of the front. An example is provided by the distribution of the Australian grasshopper *Austroicetes cruciata*, which in southern Australia sometimes reaches plague abundance. To the north its distribution is limited primarily by drought, which dries up and makes inedible the vegetation on which it feeds during its crucial immature stages. A line that includes to the south the area where the ratio of precipitation to evaporation does not fall below 0.25 during the critical month of October is approximately the northern boundary of the territory of the grasshopper. On the south edge of its range, the chief enemy appears to be infectious disease, which is fostered by dampness, particularly in September just after the eggs hatch. Here the line that includes an area with a precipitation to evaporation ratio that does not exceed 1.0

for the month of September roughly defines the southern boundary of the area occupied by the grasshopper.

The range of a species of insect is by no means fixed, since the seasonal pattern of rainfall and heat and cold varies from one year to the next. Sometimes a succession of unfavorable years will exterminate an insect over much of its range. Even aquatic insects in large permanent lakes are not immune to climatic changes. In 1953 temperature and wind conditions at the western end of Lake Erie combined to prevent normal circulation of water, with the result that a layer of oxygen-deficient water lay for a long period on the bottom of the deeper parts of the lake. Over great areas, from the islands northeast of Toledo north to the Canadian shore, mud-dwelling nymphs of the abundant mayfly *Hexagenia* were killed. Where the insects had once numbered hundreds per square meter of lake bottom, not a specimen could be taken. In this instance the rebound to the usual population level took place only the following year, either as a result of migration from shallower waters to the south, where the mayfly had continued to flourish or, more probably, by the survival of eggs that lay dormant in the mud from the year before the disaster.

Besides year-to-year fluctuations in the area occupied by a species, there may be long-term changes, in which the species continually advances or retreats over periods of decades. There are, for example, numerous species that are spreading northward in the United States. A strikingly patterned grasshopper, *Syrbula admirabilis*, previously unknown in southern Michigan, is now abundant in grassy fields near Ann Arbor. It may be that long-term climatic shifts are the cause of such changes, but it is difficult to disentangle climatic effects from the effects of widespread human alteration of the environment. Towns in the Great Plains are now miniature replicas of the eastern forests, and in the once forested regions are extensive man-made grasslands. Occupation of these new habitats by insects and other animals is slow but inevitable.

Insects constantly exert pressure on the boundaries of their area by their dispersal flights. On a summer day one may see swallows, swifts, or nighthawks hunting hour after hour for insects that are flying invisibly high in the air; the fact that whole groups of birds base their economy on the aerial insect population indicates the population to be a large one. The flights continue through the night, except that the predators are now bats that track their prey in the darkness with their echo-location apparatus. This aerial stage in the life cycle of the insect often is the dispersal phase.

Although strongly flying insects can be more or less independent of air movements, weak fliers may depend on them for wide dispersal. On hot, still days rising convection currents may be strong enough to carry small insects to heights of hundreds or thousands of feet where the strong winds of high altitudes can carry them far. Even insects that cannot fly may be dispersed by wind. Very young hairy caterpillars, such as those of the gypsy moth, may be carried like thistledown, and the eggs and minute adults of such soil insects as collembolans may be blown to dust.

Some species invade, year after year, areas that are unsuitable for their young. These migrants apparently represent a continuing wastage of the population, but, in the long run, evolutionary changes in such invaders may enable the species to colonize new areas. The small yellow butterfly *Nathalis iole* late each summer is common in Colorado, being found from the hot plains up to the tundra of the high mountains. There is as yet no evidence that this butterfly can survive the winter in Colorado, and each year's crop is assumed to migrate in from the southeast.

9

Insects and plants

About half the species of insects feed on living plant substance. Another one fourth eat decaying plant material and are usually put in a separate category called "saprophagous," but, since they often eat the living bacteria and fungi that cause the decay, many should be grouped with the plant-feeders, so that probably well over half the species base their economy on living plants.

Although each kind is usually something of a specialist, the insects taken together eat every part of the plant. Some graze on the leaves and soft stems, using heavy-toothed mandibles that nip off and grind up the plant tissues. Others, equipped with a hair-fine, hollow, piercing needle, penetrate deep into the plant tissues and draw out the nutrient fluids that circulate through the plant. Some of the smaller kinds tunnel between the upper and lower surface of a leaf or in stem or root or may grow to full size inside a single small seed.

Since living tissue must possess a fairly standard quota of proteins and other foodstuffs, one might think that one species of plant would be as good as another as food for an insect. This expectation is borne out by the catholic diet of some. The Japanese beetle, for example, feeds on about two hundred and fifty species of plants, the gypsy moth caterpillar on more than four hundred. But on the other hand, many, probably most, species refuse to eat or sicken and die when given any but one of a very few species of plants for food. Domestic silkworm caterpillars starve to death rather than eat leaves of lilac or elm. These caterpillars are raised commercially only on leaves of mulberry, although there are a few other foods that will keep them alive.

Laboratory studies show that insects may show different degrees of viability on different plant diets. When caterpillars of the nun moth are grown on leaves of apple or of various European species of oak, larch, or spruce, about 98 percent of them reach maturity, but on beech 18 percent of them die, on a pine 61 percent die, and on a species of alder 89 percent die before becoming mature.

That there may be hereditary differences among individuals in their ability to tolerate different plant foods is shown by experiments with caterpillars of a sawfly (*Pontania salicis*). When taken from its natural host, a willow, and fed another species of willow, only a small percentage of the insects survived. But offspring of the survivors showed better survival on the new willow, and eventually a strain was established that lived quite well on it.

Natural populations of insects may have the genetic flexibility needed to change from one food plant to another. Lombardy poplars that had been brought into the Fraser River area of the Pacific Northwest at first flourished but later were nearly exterminated when an insect that habitually fed on them—the satin moth (*Stilpnotia salicis*)—was accidentally introduced into the same area. With its food supply now limited, the immigrant moth began to feed on the native cottonwoods, trees rather closely related to the poplar. It is known that the poplar-reared caterpillars suffer heavy mortality when fed on cottonwood leaves, but evidently enough survived to found the race that now thrives on the cottonwoods of the Fraser River area.

Some of the most important crop pests have shifted from a native plant to a crop plant that has been introduced into their area. A classic example is the Colorado potato beetle, a black and yellow beetle that before the settlement of the West lived only in the Rocky Mountain region. Here it fed on the leaves of *Solanum rostratum*, the buffalo bur, and probably on other native solanums in the area. When the potato plant, which is a domesticated species of *Solanum*, got to the Rockies, the beetle began feeding on it and thus spread eastward, reaching Illinois in 1864 and the Atlantic coast in 1874.

Larvae of plant-eating insects are usually relatively helpless animals, and the food plant is selected by the adult female. The potato beetle identifies the host plant by its odor. Under artificial conditions, at least, the beetle may be led astray, for in the laboratory it is attracted by *Solanum demissum*, a wild potato of Peru, and will oviposit on the plant. The larva, however, after tasting the plant refuses to eat. Perhaps this error by the adult beetle is explainable on the grounds that the insect was not exposed to this species of plant when its food-finding behavior was evolving.

Why is it that the leaves of a plant are not, or can not, be eaten by a given species of insect is a complex subject. In agriculture, it is an important one, for different varieties within a single species of crop plant vary in their edibility, with some highly resistant to insect pests, others quite vulnerable. However, a proper theoretical basis for understanding the behavior of the crop pests requires studies of the immunity—or vulnerability—of plants in nature.

Ever since plants and insects have inhabited the land, there presumably has been continual evolution of new methods of attack by insects, and of new methods of defense by plants. In insects there occurred external, visible adaptations concerned with feeding—jaws and piercing beaks, for example—and hidden, internal adaptations concerned with overcoming chemical defenses put up by plants. The defenses of the plants also have both external and internal aspects—hardened armor or spines that give mechanical protection to foliage or seeds, and a variety of poisons that provide chemical defenses.

It is the chemical defenses of plants that have been longest known, although it is their effects on man, rather than on insects—the most important natural enemies of plants—that have been most studied. Indeed, some students of these chemical substances have said that they are of no use to the plant, except in that they are by-products of processes necessary for the biochemical functioning of the plant.

Most of the poisons manufactured by plants belong to a class of more or less bitter chemicals called alkaloids. For thousands of years people have known about these drugs and poisons and have shown remarkable ingenuity in discovering them. Caffeine and caffeine-like alkaloids (theine in tea, theobromine in cocoa) have been found and prized by primitive man in a variety of plants in widely scattered places. Curare (arrow poison), opium, and quinine are alkaloids. Socrates was executed with the mixture of alkaloids found in the root of the poison hemlock. The United States Department of Agriculture lists over thirty-five hundred species of plants known to contain alkaloids, containing in all about two thousand different kinds of these substances.

The role of the alkaloids in protecting plants against their natural enemies has not been well investigated, but this field of plant biology may turn out to be a vast one. Regardless of the efficacy with which these poisons turn away some potential enemies, some insects have learned to penetrate the defenses, and it is doubtful that any plant is completely immune from insect attack. Such poisonous leaves as those of the potato, tobacco, and tomato plants are the staple diet of several insects. Just how the alkaloids of these plants is tolerated is not known, although it seems likely that the insects have evolved chemical mechanisms for breaking up, or detoxifying, them. In modern times insects have evolved methods for breaking up the dangerous molecules of insecticides devised by man.

One of the factors that restricts the feeding range of insects is the inflexibility of their behavior patterns. In some the feeding mechanism is set in motion only if the taste, odor, and "feel" of plant tissue are right. Probably this kind of restriction is in part a consequence of the small size of the central nervous system of the insect. Specialized behavior is doubtless useful—in default of the trial and error and learning methods available to animals with large brains—in making certain that the insect will avoid plants that its digestive or metabolic machinery can not handle.

The kind of inflexibility that is involved can lead to errors in artificial laboratory conditions, as shown by the behavior of the bur seed fly, *Eurista aequalis*. This fly lays its eggs on the seed capsule of the cocklebur (*Xanthium*) in which the larvae feed. If the hooked spines are taken off the capsule, the fly will not lay its eggs there, although it is still suitable food for the young. If the experimenter now makes an artificial bur out of a cork studded with hooked pins, the fly gets the correct set of stimuli for laying eggs and goes on to deposit them in this hopeless situation.

A prickly pear cactus (*Opuntia inermis*) was brought to Australia in a flower pot in 1839. Cuttings from this plant were given out for hedgerows around settlers' cabins, the plant escaped and by 1925 had, in company with a few other species of introduced cacti, overrun fifty to sixty million acres in northeastern Australia. The cactus-infested land was worthless, and biologists sought some way of exterminating the plant. Neither grubbing nor burning was economical, so attention was turned to importing insects that might control the weed. Since the cactus family is native to the New World, a search was made there for enemies of the plant. Over a period of seventeen years, entomologists found about one hundred and fifty species of cactus-eating insects and mites. These were subjected to long scrutiny to make certain that they would not attack Australian crop plants and themselves become pests. Luckily the family Cactaceae is a rather isolated one botanically, and it was found that many of its enemies would eat nothing else. Eight species of insects and one of mites that became established on the Australian cacti have wrought spectacular destruction of the plant over wide areas. The most important of these biological weapons is a small moth with the apt name *Cactoblastis cactorum*.

Other plants that have been controlled to some extent by planned introduction of insects include the lantana, a thorny plant that overran parts of Hawaii, and St. John's wort, or Klamath weed (*Hypericum perforatum*), an important pest in Australia and the northwestern United States. The Klamath weed is being brought under control in the most heavily infested areas in both continents by the leaf beetle *Chrysomela gemellata*, brought from southern France.

A large part of the organic material manufactured by green plants is fashioned into wood. Insects have shown their dietetic versatility by exploiting this massive but refractory source of nutrients.

The chief constituent of wood, amounting to about half, is cellulose. Although apparently none of the foliage-eating insects are able to digest the cellulose of leaves, many of the wood-eating insects are able to digest cellulose and rely on it as a major source of energy. Some beetle larvae secrete digestive enzymes that break down cellulose, but other wood-eaters, such as the termites, have in their digestive tracts large quantities of living microorganisms that digest the cellulose. The nutrient-rich excretory products of these microorganisms are then absorbed through the gut of the host animal.

Lignin, which comprises from 18 to 38 percent of wood is, so far as known, not utilized by any insect. Starches and sugars, which are easily digested, are concentrated in the sap wood, where they may amount to as much as 10 percent of the dry weight of the wood. Some insects, however, rely on the scarce sugars and starch of the heartwood and, being unable to digest cellulose, must devour great quantities of wood to live.

There is a small amount (1–2 percent) of protein in wood, and this provides the nitrogen for protein synthesis by the wood-eating insect; the once-held belief that the symbiotic microorganisms in the digestive tract of many wood-eaters must be able to fix atmospheric nitrogen to supply the nitrogen needed by the insect is probably not correct. Decaying wood may be further enriched by nitrogen-containing compounds brought up from the soil by fungi.

There are two groups of insects—the gall midges (family Cecido-myidae) and the gall wasps (Cynipidae)—totaling some thousands of species, that are specialists in forcing plants to alter normal growth so as to provide both food and shelter. These two families are not related, the first being in the order Diptera, the other in the Hymen-optera, and the ability to produce galls also has appeared independently on a minor scale in many other groups of insects.

Some galls are relatively structureless swellings, but others, and especially those produced by the gall midges and wasps, are as precisely ordered and elaborate as buds or flowers. One type of gall easy to find in winter is that on the tips of willow stems. These symmetrical, cone-like structures are usually taken for normal parts of the willow, but each houses larvae of gall midges whose presence is necessary for gall formation. Spherical galls, sometimes ornamented with spines or woolly pubescence, that are common on twigs and leaves of oaks, are produced by gall wasps.

The structure of the gall is as characteristic as the morphology of the insect, and in many instances it is easier to identify the gall-maker from a specimen of the gall than from the insect itself.

After the larva of the gall insect hatches from the egg that has been inserted in the plant tissue, it secretes, presumably from its salivary glands, the substance or substances that induce gall formation. The surrounding plant tissues are irritated or stimulated in such a way that they grow, not in the way normal in the development of the plant but in such a way as to produce the gall characteristic of the insect. Plant cells near the insect become large and multinucleate and show abnormalities of cell division; these form the nutritive layer. Cells that are distant are usually more nearly normal but become organized into tissues that form the hard protective layers of the gall. It is obvious that the substances secreted by the larvae act in a manner analogous to that of the chemical organizers of embryological development, since their presence regularly induces the formation of definite structures.

Since gall structure is determined primarily by the insect, on the leaves of a single oak, for example, there may be galls of quite differing appearance, each produced by the appropriate gall wasp. Also, if a group of closely related gall insects infest several unrelated species of plants, their galls tend to be alike, in spite of the diversity of plants on which they develop.

The tumor-like plant galls are in some ways reminiscent of the malignant tumors of animals, and it is not surprising that they have been studied with the idea that principles relevant to cancer formation in animals might be uncovered. An important difference between malignant tumors and plant galls is that the growth of the former appears to be self-perpetuating once begun, whereas the development of the gall depends on continued secretion of the gall-inducing substance.

The foregoing emphasizes the prey-predator aspect of the relationship between insects and plants; there is, in addition, the mutually helpful, or symbiotic, one in which insects carry pollen, thus crossfertilizing the plants, with the plants on their side contributing nectar and some of the pollen to the insects.

The relationship between insects and plants is a curiously close one.

Perhaps half the kinds of insects now alive, and by far the greater bulk of them, live on plant substance. But it is a two-way road, for half or more of the species of plants depend on insects for sexual reproduction. When plants lived only in the water, the male gamete or sperm, capable of swimming, could transport hereditary material, but, with the invasion of the land, the motionless plants had to exercise a good deal of evolutionary ingenuity to keep sex alive.

Basically, the solution hit upon was the development of tiny male plants, called pollen grains, that were light enough to drift in air currents. It is obvious that enormous numbers of them must be produced by a plant if, on the average, even one grain is to alight on the receptive surface (the stigma) of another plant. Thus, the great coniferous forests shed tons of dust-fine pollen into the air, with most of it going to waste on barren ground or giving people hay fever.

By the time the extensive plant cover of the Coal Age (Carboniferous) had developed, there were winged insects in existence capable of transporting pollen, and the astonishing capabilities of evolution for taking advantage of available opportunities are shown by the varied ways in which this means of transportation was exploited.

First steps in enlisting the winged insects into the pollen-transporting enterprise are not difficult to imagine. The huge, protein-rich pollen masses of the early wind-pollinated plants must have been an attractive source of food for many insects, and, as they moved from plant to plant, they carried pollen. Variants that were more attractive to insects would transfer pollen more accurately to other plants. An early attractant, besides the nutritious pollen, was doubtless sugar water, or nectar. Add to the nectar a visual advertisement—bright color and recognizable pattern—and the flower has evolved, a flower being in essence a receptacle that offers nectar in such a way that the insect drinking it brushes against the pollen. Some flowers offer pollen only, and, as a rule, even the nectar flower sacrifices some of its pollen to the appetite of the insect visitors.

Nectar is a watery solution of sucrose (ordinary table sugar) and of its constituent sugars, glucose and fructose, in varying proportions. The concentration of sugar in nectar may be as high as seventy percent by weight. Usually nectar is secreted by definite structures, the nectaries, situated near the base of the flower. A single flower of moderate size may secrete from less than a milligram to as much as twenty-five milligrams of sugar a day. On flowery meadows this means that several tens of grams of sugar an acre are produced daily. If one is quiet on such a meadow on a warm day, one gradually becomes aware of an all-pervading hum, that of tens of thousands of flying insects. Probably most of this flight is powered by nectar. The amounts of energy involved make possible a considerable amount of insect flight: it can be calculated that the yearly sugar production of a single acre of meadowland would power something like a half-billion miles of flight by the small fruitfly, *Drosophila*.

There is some reason for thinking that beetles were among the first pollinators, since beetles regularly pollinate some of the most primitive flowering plants. But the pollinators extraordinary are the twenty thousand or so species of bees. All bees, both larvae and adult, feed primarily on nectar and pollen. In their closely intertwined histories, bees and flowers have mutually shaped each other's evolutionary development: not only have the bees become adapted in intricate and varied ways to collect pollen and nectar from plants, but flowers also have become adapted to the insects. The pleasing forms of flowers are "meant" primarily for the eyes and bodies of bees.

10

Insects and other animals

The relationship between insects and other animals is predaceous both ways: insects feed on other animals, and other animals prey on them. But the insects come out second best, providing much more food for the rest of the animal world than they get back in return.

Whole groups of vertebrate animals base their economy on insects. Stream fishes depend not only on the abundant aquatic insect life but also on the insects that fall into the water from streamside vegetation. Those most abundant amphibians, the frogs and toads, are mainly insect eaters; a toad kept in captivity for nearly sixteen years was fed on cockroaches, and during this time it ate an estimated seventy-two thousand of these insects. Among the reptiles, most of the smaller lizards are insect-eaters. The smaller birds depend much on insects, even the grain-eaters often raising their young on a protein-rich insect diet. Birds such as the nighthawks, swifts, and swallows are so completely adapted for catching insects on the wing that they can feed in no other way. Woodpeckers have a complex of adaptations that enables them to capture insects hidden under bark or in wood. The water ousels are birds of mountain streams adapted to walk and fly under water in pursuit of aquatic insects. Among the mammals, the bats of northern regions feed only on insects caught on the wing. Shrews, abundant but little-seen mammals of the forest floor, are in large part insect-eaters. Certain strange and ancient mammals of the southern hemisphere and Asia live on termites: the anteaters, the aardvark, the echidnas, and the pangolins. Even such large mammals as bears depend to some extent on insects for food, and man did and still does eat them.

So closely and continually have insects been hunted by the keen-eyed birds that one of their characteristic defenses has been to make themselves as nearly invisible as possible, with the result that among the insects are found the most remarkable examples of camouflage in the animal world.

A dozen or so graduate students in biology were shown a square-foot area on the trunk of a black walnut tree and were asked if they could find the three caterpillars, each over an inch long, that were resting there. The only rule of the game was not to touch the tree—they could look until they gave up. So closely did the caterpillars match the color of the bark that some students failed to find them. These insects were larvae of the large underwing moth (*Catocala*). They lie quietly on the bark during the day, helplessly exposed to their enemies, then move out at night to the ends of branches to eat leaves.

Another caterpillar that is difficult to see is the tomato worm, the larva of a sphinx moth. One of these huge, finger-sized caterpillars can half destroy a plant overnight, but the gardener might search half an hour before he finds the culprit clinging quietly to a stem. The color exactly matches the green of the plant, except for a row of diagonal light stripes. These, instead of giving the caterpillar away, break up the cylindrical outline one is looking for to make it even more nearly invisible.

Concealment can be achieved by imitation of the form of inanimate objects. The impressive walkingstick insects are usually formed either like slender twigs or else like stems and leaves. Although some are the longest insects living—a foot in length, not counting the long, slender legs—they are almost impossible to detect in life in their natural surroundings, and even a dead specimen looks like a scrap of vegetation. The color as well as form of these cryptic insects is concealing, but there are some walkingsticks that are stout-bodied, little resembling twigs, and are conspicuously black. These boldly colored species emit evil-smelling and caustic secretions that protect them.

In the early 1900s, discoveries in genetics, especially the discovery of the role of the chromosomes in governing heredity and variation, made the theory of evolution by natural selection more or less unfashionable, and for a time the survival value of such things as the leaf- or twig-like form and color of the walkingsticks was called into question. Critics of selection theory pointed out that these presumably protected insects got eaten anyway. Later, opinion swung the other way for a variety of reasons, the most important being the realization that not absolute protection or survival but improved protection or chances of survival were important in producing evolutionary change. The fallacy in the argument that the walkingsticks get eaten in spite of their presumed protection may be seen by looking at an analogous situation in human affairs. Even a heavily armored army tank is sometimes pierced by certain kinds of shells, but this does not mean that the armor is therefore of no use and might as well be discarded.

Another kind of protection that has evolved in connection with the presence of such sharp-eyed predators as birds is one called mimicry. Here the insect looks like another insect that has good aggressive protection—poison, an evil-tasting chemical, or a fiery sting. The dangerous or distasteful insect is usually boldly colored, often red or yellow and black, with the result that it is easily recognized by a would-be predator. These genuinely unsuitable insects are mimicked by a host of insects resembling them in details of form, color, and movement but fraudulent in that they do not have the undesirable qualities of their models.

Various examples of mimicry will be given in later discussions of the different insect groups, but one will be given here to show the principle involved.

The two-winged flies of the family Syrphidae (flower flies) spend a good deal of time on flowers, where they drink nectar and eat pollen. Here they are in company with bees and wasps, which are protected to some extent against predation by their stings. Many of the syrphid flies have the yellow and black striping of wasps. The resemblance to bees or wasps may be rather vague and general, or it may be astonishingly precise. The syrphid fly *Eristalis tenax*, or drone fly, closely resembles the honeybee. It might be difficult to observe predators in the wild being deceived by this resemblance, but the effect of the mimicry on students in beginning entomology is obvious enough. Almost invariably they are extremely circumspect in dealing with the harmless fly, and it takes several hours of experience in collecting flower insects to learn to distinguish at a glance between mimic and model.

Another syrphid mimics a bumblebee. In the Colorado Rockies in the late summer there is a common bumblebee that is made conspicuous by a band of red hair across the back of its abdomen. On the same flowers at the same time there is occasionally seen a species of a hairy syrphid fly (*Volucella*) with a band of red hair in the same position. The comparative scarcity of the mimic is, of course, necessary if the scheme is to work. If the really dangerous bee is the exception rather than the rule, then the predator might find it difficult or not worthwhile to learn to avoid the warning bee coloration.

Insects are themselves of such small size that they do not feed on the vertebrate animals by predation but rather by stealth, drawing small amounts of blood or living inside the bodies of their victims as parasites.

The most familiar blood-drinkers are such two-winged flies as the fiercely biting mosquitoes and black gnats, which are incredibly abundant in the wet lands of the far north, and the swift deerflies and horseflies (Tabanidae) that swarm around livestock. Birds and mammals, including most human beings alive today, are infested with the blood-feeding lice and flies. It is their habit of feeding on blood that has caused insects to bring about the deaths of millions of human beings. The microorganisms that cause malaria and yellow fever have no way of getting to a new victim unless carried by mosquitoes, and African sleeping sickness is carried only by tsetse flies.

Insect parasites that live inside larger animals also are Diptera. Inside the digestive tract of horses, and absorbing a share of the nutrients there, are larvae of the botflies (Gasterophilidae). Larvae of warble flies (Hypodermatidae) wander through the body of cattle, burrowing over the surfaces of the intestine and other internal organs without causing serious damage. However, when the larvae mature, they leave by digging out through the skin of the back, which makes open wounds and spoils the hide for leather.

11

Insects versus insects

Insects' worst enemies are themselves. Of the nearly one quarter of a million known species that eat animals instead of plants, most feed on other insects. Probably every species of insect in existence has one or more insect enemies.

The insects that feed on other insects can be classified into two groups on the basis of their habits: the predators and the parasites. In practice it is not easy to draw the line between the two, but in general the predator kills more than one victim during its life, whereas the parasite usually has only one. The parasite often lives inside its victim and usually kills it. Some parasites do not kill directly but appropriate the food store that the mother has laid up for her offspring, causing its death by starvation.

The predatory stratagems of the vertebrates are thin and monotonous affairs compared to the miniature but varied ferocity of the insect world. Some insect predators rely on dash and speed. Such are the large dragonflies that spend most of their adult lives patrolling the air in search of insects, which they catch and devour on the wing. Robber flies (Asilidae) sally out from high vantage points to capture insects on the wing or on the ground. The tiger beetles (Cicindelidae) are sharp-jawed insects whose long legs carry them swiftly over the ground after their victims.

Other predators are rather sluggish insects that wait for their prey to come within reach. The heavily-armored ambush bugs (Phymatidae) (Fig. 48) lie on flowers, more or less concealed by camouflage, and seize insects that come for nectar and pollen. The young of the myrmelionids and of certain two-winged flies dig pitfalls in sand or dust and wait at the bottom for their victims to tumble in. Maggots of some fungus gnats (Mycetophilidae) spin silken webs, which they deck with droplets of oxalic acid that kill insects touching them.

The "tooth and claw" of predatory insects are most usually the mandibles, fashioned into sharp blades, and the front legs, which have strong muscles and sharp spines that fit them for grasping and holding. Widely different, unrelated groups have independently evolved raptorial front legs of this kind: the praying mantids (Orthoptera), the ambush bugs and assassin bugs (Hemiptera) (Fig. 30), and the mantispids (Neuroptera). Occurring in some of the Hemiptera is a variant type in which the legs are equipped with non-skid, velvety gripping surfaces instead of spines.

The cutting mandibles are used to kill and cut up the prey. Some predators, however, operate on a different principle and have the mandibles or other mouthparts fashioned into a stabbing beak, used first to inject poisons or digestive enzymes into the victim and then to suck out the blood or digested tissues. The predatory Hemiptera just mentioned have such mouthparts, as do the robber flies.

Another weapon useful in both offense and defense is the sting of the bees and wasps (Hymenoptera). Whereas the sting of the bee is purely a defensive weapon, that of the wasps is used to kill or stun prey. The sting, which occurs only in the Hymenoptera, is a modification of the egg-laying device, or ovipositor, at the tip of the abdomen. Since it is usually equipped with poison glands, its effect is more than of a simple puncture.

The insects that are parasites of other insects far outnumber the predators, both in individuals and species. Although of great economic importance, so diverse and obscure are they (many are of almost microscopic size, growing to maturity inside a single egg of the host) that most species still remain undiscovered. Most parasites belong to two orders: Hymenoptera and Diptera. It is the helpless larva, nearly immobile, that is the parasite, with the responsibility for finding the

FIG. 30. *The spiny front legs of this predatory assassin bug are used in seizing prey. The jointed structure under the head is the sheath that protects the hair-thin piercing beak that sucks out the body contents of the victim.*

host falling upon the active, far-ranging adult. The most important groups of Hymenoptera that are parasites are the ichneumon wasps and the tiny chalcid and braconid wasps; important among the parasitic Diptera are the tachinid flies.

Details of the habits of various groups of parasites will be described in later chapters where these groups are taken up individually. In general, no stage in the life history of insects is free from attack by parasites. The eggs are attacked by minute species. The larval stage of the host may be large enough to house a single large parasite or hundreds, or even thousands, of small parasites. When the parasites are feeding inside the body of their growing host, typically they at first avoid the vital organs, allowing the victim to feed as long as possible before they finally kill it and themselves become mature. A favorite stage for attack for many parasites is the immobile pupa. If this is protected by a cocoon, the parasite has a long egg-laying tube that penetrates the shelter and reaches into the pupa. And, finally, the adult stage serves as the host for some parasites. The parasite may then have to display the dash of the predator. Certain grasshoppers, for example, are parasitized by flies of the family Sarcophagidae. When at rest, the grasshopper is protected by the thickened front wings that are held like shields over the body. The parasite attacks only when the grasshopper is in flight. As the grasshopper rises from the ground, the much smaller fly darts in, like a fighter plane attacking a bomber, to place its maggots on the exposed hind wings of the grasshopper. From here the maggots burrow into the body of their host.

As a rule the adult parasite deposits eggs in or on the victim, but in some of the Diptera, as in the case of the sarcophagid fly described above, the young are born alive and are placed on or near the host.

A very few of the parasites of other insects attack their hosts in the manner that a mosquito attacks a man. Some of the tiny flies of the Ceratopogonidae bite through the skin and suck the blood of other larger insects; some, it is said, even bite mosquitoes.

Every long-established and abundant insect supports other species that parasitize it or prey upon it. It may be freed of these enemies when it invades new areas, but the new-found immunity does not last long. Sooner or later, its old enemies find it out, or some of the thousands of species of insects that surround it learn to exploit it.

The insect pests that man has introduced into new areas may thrive mightily, at least for the time being, because their community of enemies has been left behind. One of the most intricate and challenging problems taken up by biologists is that of controlling such destructive species by pitting other insects against them. Although there have been some individual successes over serious pests, this field of applied entomology—the biological control of insects—is still in its infancy, and much reliance is placed on chemical control, the use of insecticides. Crude and widespread use of insecticides in the long run creates more problems than it solves, and the need is for careful studies of the biology of each situation, so that the insecticide can be employed as a precision instrument that supplements the varied methods of biological control.

The obvious starting point in controlling a destructive introduced insect is to explore the country of origin for insect predators and parasites. Entomologists of the United States Department of Agriculture have made world-wide explorations for more than half a century and have sent back for study in the laboratory thousands of species of potentially useful insects. A spectacular early success was the control of the cottony cushion scale of orchard trees in California by the Vedalia (*Rodolia cardinalis*), a ladybird beetle from Australia.

Exotic parasitic or predatory insects are introduced most easily into areas with a relatively equable climate. It is in Southern California and the Hawaiian Islands that biological control has been outstandingly successful. Probably the variable and rigorous climate in continental areas is the main barrier to introducing beneficial insects. So far, entomologists have relied mostly on the relatively primitive method of hunting to get their insects for biological control; that is, they take nature as they find it. It seems reasonable that the next and most complex as well as successful stage in the science will be one in which entomologists breed varieties of parasites and predators that are especially well adapted to local environmental conditions.

When one thinks of the intricate field and laboratory investigations needed for this kind of delicate manipulation of the biological world—

the use of insects to control other insects or the breeding of varieties of crops resistant to insect attack—it becomes apparent that technological unemployment is not an absolute consequence of scientific advance. It would take a large percentage of the population of a highly advanced country to carry out adequate research on living materials, not only in the problem of the control of noxious insects, relatively only of minor interest in the whole range of human activity, but also in other areas of biological research.

12

Insect life in the water

If one watches quietly beside a forest pond, one gradually becomes aware of the teeming insect life beneath the surface. If early in the spring, there will be swarms of mosquito wrigglers, looping with extravagant motion and little speed through the water. Streamlined beetles cruise about powerfully, now and then halting to float to the surface to replenish their air supply, and Hemiptera swim expertly with long, oar-like legs. Other, less conspicuous insects live in the debris on the bottom and have to be dredged out to be seen.

These insects are descendants of land-dwellers, for they have air-breathing respiratory systems which have been modified to get oxygen out of the water. The invasion of the fresh waters probably began a very long time ago in the history of the insects, since some of the oldest known winged types, which resembled living dragonflies and mayflies, found fossil in Carboniferous rocks are believed to have had aquatic young.

The stoneflies and the true dragonflies and mayflies are the most primitive aquatic insects alive today, having been found as fossils as far back as late Paleozoic times. It is the young of these that are aquatic, and the adults are flying insects that return to the water only to lay their eggs on the surface. The young are well adapted for life in the water, usually breathing by means of a more or less elaborate set of tracheal gills.

Tracheal gills usually are flattened plates or long filaments that offer a large absorptive surface to the water. The gills are filled with blood and also contain fine branches of the tracheal system. Into these fine air tubes oxygen diffuses from the water and is then distributed to the rest of the body by way of the main part of the tracheal system. When the adult emerges to take up its aerial existence, it sheds the tracheal gills, the spiracles open, and air can go directly in to the tracheae.

With some minor exceptions, the only insects that live in the water as adults are some of the Coleoptera and Hemiptera. They have the ability to fold the wings back over the abdomen to give a streamlined body form, and the more delicate hind wings are shielded by the heavy front ones. The young of the aquatic beetles and Hemiptera also are aquatic, but the whole life cycle is not necessarily spent in the water, since most of the adults are good fliers and in dispersal flights spread the species far and wide into suitable habitats.

These adult insects breathe, not by means of tracheal gills, but with artificial gills that they make themselves. A diving beetle (Dytiscidae) carries a bubble of air under the wing covers. This bubble is given further stability by a layer of felt-like hairs on the top of the abdomen, and it connects with the tracheal system by way of spiracles on the abdomen. Occasionally the beetle comes to the top, protrudes the last two spiracles out on the water surface, and takes into the tracheal system a supply of air. Also it renews the large air bubble. If the oxygen supply taken in is measured, it will be found that it is enough to enable the beetle to stay under the water for a limited time only. But if, say, the oxygen taken in is sufficient for half an hour of life, the beetle in reality is able to stay under water for several hours. This is possible because the bubble serves as a gill, oxygen from the water diffusing in through the membrane of the bubble as fast as it is used up by the insect. In spite of this, the bubble dwindles away because the nitrogen of the bubble slowly leaks out into the water, and it is this that makes it necessary for the beetle eventually to come to the surface.

Larvae of many of the two-winged flies (Diptera) are aquatic. These legless creatures are not good swimmers and would not seem to be in line for designation as aquatic insects *par excellence*, but they are in numbers of species and sheer bulk among the more important members of the aquatic insect world. The mosquitoes and their relatives, the small midges (Chironomidae), are extremely abundant, and the latter probably rank as the most abundant and ubiquitous of the insects of the water.

Mosquito larvae, or wrigglers, spend most of their lives just below the surface, so that, although aquatic, they have ready access to air, which they get by putting a short breathing tube out on the surface film. When frightened, they swim downward with their characteristic looping contortions, and the store of air carried in the tracheal system may last as long as ten minutes. The opening of the air tube is guarded by water-repellent hairs or flaps. The old method of killing mosquito larvae by spreading oil on the water works because the waterproof waxes on the air tube attract rather than repel the oil, which then clogs the tracheae and suffocates the insect. Some mosquito larvae get oxygen by driving their sharp breathing tube into the tissues of water plants.

Midge larvae are not dependent on the surface and breathe by absorbing oxygen directly from the water through their thin skin. Midge larvae that live in deep water or in mud where oxygen is scarce have hemoglobin dissolved in the blood. The red, oxygen-carrying pigment of these so-called blood worms is basically like the blood pigment of the vertebrate animals. Midge larvae abound in a great variety of aquatic habitats, even living in the bottom of deep lakes, and in coastal waters of the sea to depths as great as one hundred feet. Even the adult of one of these marine midges stays below the surface.

The air-tube method of breathing has been highly developed by the larvae of some of the syrphid flies. These "rat-tailed maggots," themselves about an inch long, have an extensible tail that can be stretched as far as six inches to the surface. This arrangement is perfectly adjustable, and the insect retracts or extends it to suit the depth.

Further details or habits and structure of the groups that live in water will be found in the chapters on the various orders. A summary of the groups of aquatic insects follows.

A. Groups in which only the young are aquatic
 1. All members (except a few aberrant types) of the orders Ephemerida (mayflies), Odonata (dragonflies and damselflies), Plecoptera (stoneflies), and Trichoptera (caddis flies)
 2. Several families of Diptera (two-winged flies) and Neuroptera
 3. A few moths (Lepidoptera)
B. Groups in which both young and adults are aquatic
 Several families of Hemiptera (true bugs) and of Coleoptera (beetles)
C. Groups which are not really aquatic but which are adapted for life on the surface of the water, some even living on the surface of the open ocean
 Principally the family Gerridae in the Hemiptera

Part four

Parade of the insects

13

Insects without wings

In our summary of arthropod evolution, it was seen that there were two main lines: (1) the chelicerate line, including the spiders, in which the first pair of leg-like appendages were changed into chelicerae, and (2) the mandibulate line, with crustaceans, myriapods, and insects, in which the first pair were either kept to form an extra pair of antennae or were discarded, and the second pair changed into mandibles.

An important phase of the history of the mandibulate line was the progressive loss of walking legs down to a minimum of three pairs. According to the definition of insect used here, the insects appeared when this line reached the six-legged stage.

We do not know when this happened, but the oldest known fossil insect (the wingless *Rhyniella praecursor*) is from the Devonian and is about 300 million years old. Insects may have been in existence long before this.

These early insects, of course, did not have wings. Although there are few fossils recording this early wingless phase of insect history, there are several living groups which are believed to represent it. They are called primitively wingless, to distinguish them from other more modern types that, like worker ants, are clearly related to and descended from insects that do or did have wings.

There are several anatomical features that serve to connect these primitively wingless insects with their myriapod ancestors. Some retain on the abdomen the vestiges of several extra pairs of legs. Others add a body segment at each molt, like the myriapods, but unlike the winged insects. And, finally, some have antennae with muscles inside for moving the segments, again like the myriapods but unlike winged insects.

These wingless insects are usually grouped together into a subclass called the Apterygota, which stands in contrast to the subclass of winged insects, the Pterygota. In the Apterygota are four orders: Protura, Collembola, Diplura, and Thysanura. These ancient groups differ profoundly from each other, and also from the winged insects, and this difference has caused some authorities to regard one or two of these orders—the Protura and Collembola—not as insects at all, but to be ranked as separate classes equal in standing to the class Insecta.

Nearly all of the existing primitively wingless types live in dark and damp places—in soil, turf, rotten wood, under stones or loose bark. In the soil they are part of a complex community of small animals that live on dead vegetation that falls from the lighted zone of green plants above, or on each other.

Being adapted to life in damp environments, they dry out quickly when exposed, and their sensitivity to dryness is taken advantage of to collect them from the soil. A handful of soil may be placed in a funnel closed at the bottom with a screen. An electric light bulb over the funnel will gradually dry out the dirt, causing the animals to move downward and to fall through the screen into a collecting jar below. This apparatus is the Berlese, or Tullgren, funnel.

The Protura are minute insects, none more than two millimeters long, and escaped notice until 1907 (Fig. 31). However, they are not rare, and specialists have learned how to find them in numbers, so that new species will be added to the hundred or so now known, and more will be learned of their biology.

They walk slowly on only four legs, using the first pair for antennae. The short, piercing mouthparts are withdrawn into the pear-shaped head. Although there are no recognizable eyes, there is a pair of small structures on the head that may be light-sensitive, which would serve to keep these delicate, moisture-requiring animals from wandering out into the light.

The proturans are of theoretical importance because they are the least insect-like of the six-legged arthropods, the main points of interest being the absence of antennae, the addition of new body segments after hatching, and, internally, the absence of Malpighian tubules. Some species of these minute animals are without a tracheal system and breathe through the skin. Those that have the system have only two pairs of spiracles, each with an isolated set of tracheal tubes.

The Collembola, or springtails, are by far the most prevalent of the primitively wingless insects. They abound in the soil; they are common under stones, fallen timber, or loose bark; in early spring they swarm on ponds, where they live on the surface film of the water; some live on the surface of snow. There are more than a thousand species.

Most are only one or two millimeters long. Collembolans are well represented on all continents, and their distribution is remarkable in that some species may be nearly world-wide. One, for example, is found in Japan, Tierra del Fuego, North America, and Europe. Perhaps these cosmopolitan species may be so ancient that they had adequate opportunity to work their way over the land bridges that at one time or another connected the continents.

The oldest fossil insect, the Devonian *Rhyniella* already mentioned, is a collembolan. These minute fossils, only a millimeter long, were described in the 1920s from chert rocks that had long been known for

FIG. 31. *Order Protura.*

the plant fossils in them. Only the heads were preserved in the first specimens discovered, but they were thought to be collembolans. This was confirmed by a few later finds, which showed a thorax, with three pairs of legs, and a legless abdomen. The structure of the abdomen was something like that of living collembolans, but the tip of the abdomen is missing in all specimens, so it is not known if they had the springing "tail." No other fossil collembolans are known through some hundreds of millions of years until Cretaceous times. In Cretaceous amber from Manitoba a well-preserved collembolan has been found that has the characteristic structures for leaping. This fragmentary fossil record indicates that the collembolans were well differentiated far back in the history of the insects.

Many collembolans live in exposed places, as on vegetation. Perhaps correlated with the ability to live in dry air is the hairy or scaly skin, which is more or less waterproof. The scales, which are flattened hairs, are much like those of butterfly wings. Like the butterfly scales, they are sometimes pigmented, giving to collembolans alone of the wingless insects a bright color pattern.

The antennae are always few-segmented, and like those of the myriapods, have internal muscles that move the segments. Just behind each antenna is a ring-shaped sensory organ of unknown function. The eyes are coarse-grained clusters of eight or fewer separate elements that obviously give only poor definition. In some the mandibles are missing, in others they are well developed, with a specialized grinding or molar plate. The fact that the abdomen has only six segments, even in the embryo, distinguishes the collembolans from other insects, which have ten to twelve abdominal segments, at least as embryos.

A very characteristic collembolan structure is the collophore (meaning glue-bar), which has been considered to be an adhesive organ for holding the animal to smooth surfaces and which gives the order its name (Fig. 32). The structure is usually kept withdrawn into the abdomen but can be thrust out through a short ventral tube on the first abdominal segment. When out, its surface is moistened by a secretion from the salivary glands that is carried back by a groove on the underside of the head and thorax. The most recent theory of its function is that it is used for breathing at times when the general body surface is too dry for skin-breathing. Most collembolans are without a tracheal system.

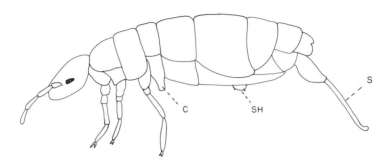

FIG. 32. *Order Collembola. C, collophore; S, spring; SH, spring holder.*

Most species bear on the third and fourth segments of the abdomen two pairs of modified legs. The hind pair is longer and constitutes the spring. When not in use it is folded forward under the abdomen and is held in place by the first pair, which form the spring holder. In jumping, the spring is released from its catch and strikes the ground. When these minute animals jump, they seem to disappear, and the observer who finds a patch of them on an overturned log is impressed by the way the group dwindles away as if it were evaporating.

These insects normally feed on plant materials. Those of the surface of spring ponds graze on algae, the snow fleas on accumulated wind-blown pollen; several species feed above ground on green vegetation and damage crops.

Members of the order Diplura are larger than the other apterygotes, some reaching a length of nearly two inches. They have some of the aberrations of the preceding orders: the antennae have internal or "intrinsic" muscles, the mouthparts are withdrawn into the head, and there are no Malpighian tubes. A typical insectan structure is the pair of cerci, antenna-like organs sensitive to touch and to vibrations in the ground that are situated at the tip of the abdomen (Fig. 33). All are eyeless.

This order, with some two hundred species, includes both herbivores and carnivores. One group has the cerci transformed into a pair of stout pincers, used to capture other small animals.

The order Thysanura is, of the primitively wingless groups, most like the winged insects.

One of the few fossil apterygotes known is a thysanuran, of quite modern appearance, from the Triassic of the Urals; undoubtedly the order is much older than this. There are about four hundred living species, found on all continents.

In those species that live in dry environments, the integument is covered with scales, often silvery in color. There are four main points of resemblance between the thysanurans and the winged insects: the antennae, which are without intrinsic muscles; the location of the mouthparts, which are on instead of in the head; the mandibles, which in the more advanced groups have two points of articulation with the head; and the eyes, large and with numerous visual elements (Fig. 34). There is a long median filament between the cerci. The triple set of caudal filaments is found elsewhere only among the may-flies, which are among the more primitive winged insects.

Many of the thysanurans venture into dry situations. The firebrat

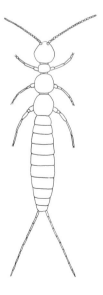

FIG. 33. *Order Diplura.*

(*Thermobia*) frequents bakery ovens, and the fast, shining silverfish (*Lepisma*) is a common household insect. It may be that in Paleozoic times there were many species of insects of this type living in open, dry habitats, but, if so, they have been superseded by the modern winged groups.

In all winged insects, growth and molting cease at sexual maturity. Among the thysanurans, however, both continue after maturity. The female of *Thermobia* may double in size after beginning to reproduce and molts up to thirty times in the ensuing two years of her life. She commonly lays a batch of eggs after each molt and mates each time.

FIG. 34. *Order Thysanura.*

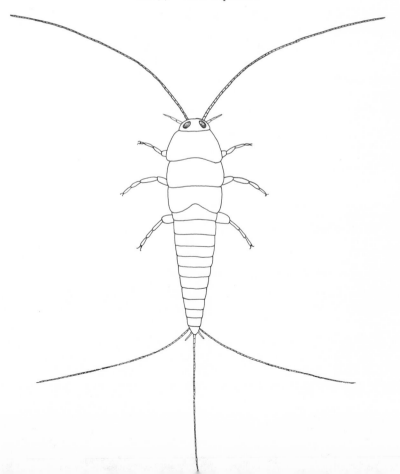

These primitive insects have complicated mating procedures, and the method of fertilization in *Thermobia* is without parallel among the winged insects. The male attracts the attention of the female by approaching several times, wagging his head from side to side, and touching hers. After a time, the approaches are varied as he whirls about, abdomen upturned and twisted to one side, and walks past her brushing against her legs. The female responds to this contact by moving forward a short distance, then turning around. After half an hour or so of these preliminaries, the male deposits a single spermatophore on the ground a short distance ahead of the female and dashes past her with unusual speed, again brushing against her legs. This causes her to move forward, this time over the spermatophore, which is seized by her genitalia.

14

Primitive winged insects

In an account of winged insects it has been traditional to take up the Orthoptera first because of all the living winged orders they seem to be most like the thysanurans. In 1928, however, the Russian student of fossil insects A. B. Martynov and the American insect anatomist G. C. Crampton independently stated their conviction that the Ephemerida and Odonata were necessarily the most primitive living winged insects because they lacked a mechanism for flexing or folding the wings out of the way when not in use, a mechanism present in most other winged insects. They proposed that the first flying insects must have had wings that could not be flexed and that the flexing device must have originated long after flight appeared. As living representatives of the earliest stage in the evolution of winged insects, the Ephemerida and Odonata, then, should be ranked as the most primitive living orders. They were put in a group called by Martynov the Palaeoptera.

For various reasons, however, the traditional approach is preferable here, and, with this brief reference to the Palaeoptera-Neoptera hypothesis, we will proceed to a discussion of the Orthoptera.

The Orthoptera are the insectan equivalent of the grazing mammals, since they are abundant (especially in grasslands), are large, and are voracious feeders on vegetation. The Orthoptera are important as primary converters of plant into animal substances and thus are an important source of food for predatory animals.

The fossil record of the Orthoptera and of their near relatives is a complicated one and extends back to the beginning of the known history of flying insects. The oldest known fossils of winged insects consist of three wings, two of them fragmentary, from Germany, Czechoslovakia, and Pennsylvania, in rocks of the Upper Carboniferous (Pennsylvanian) period. One of these wings is orthopteran-like and is assigned to a closely related, more primitive order, the Protorthoptera. In rocks of the same period, but a few million years later, fossil insects are more abundant, and there are many cockroaches not much different from living ones. The cockroaches have therefore existed for at least 240 million years. True Orthoptera other than cockroaches do not appear in the fossil record until the beginning of the Mesozoic, about 50 million years later, but, since the Protorthoptera grade into the Orthoptera, it is not possible to state with certainty when their record does begin.

The Protorthoptera were a diverse group that lived through Upper Carboniferous and Permian times, contemporaneously with the ancient cockroaches. Some of the early Protorthoptera had, in addition to the two pairs of large flying wings, a pair of short, rigid lateral expansions on the prothorax. This ancient group underwent a diversification similar to that of the later Orthoptera, for some had the front legs modified for seizing prey, like the modern mantids, and others had hind legs modified for leaping.

Living Orthoptera differ from the extinct Protorthoptera in lacking the short third pair of "wings" on the prothorax, although the wide flanges on the prothorax of some living cockroaches look suspiciously like them. The main flight surface of the orthopteran is the hind wing, very wide and almost semicircular, and pleated like a Japanese fan (Fig. 35). The wing folds along the pleats when it is at rest, and fits under the narrow, somewhat thickened front wing. The front wing, however, does contribute to flight, since it is powered by muscles of respectable size. There is no mechanism for coupling the two pairs of wings.

The mouthparts of living Orthoptera are what would be expected for a primitive insect; they lack the varied specializations of most other insect groups and are like those of the thysanurans.

The twenty thousand species of Orthoptera are grouped into five suborders: Blattaria—cockroaches; Saltatoria—grasshoppers (locusts) and crickets; Mantodea—praying mantids; Phasmodea—stick and leaf insects; and Grylloblattodea—ice crickets. These show different degrees of relationship to each other, and some authorities segregate one or more out as distinct orders, but all are genuinely related, and there is considerable justification for following the conservative course of lumping them into a single order.

The flat, wide-bodied, fleet cockroaches have existed for well over 200 million years without essential change. They are the most abundant fossil insects in the Carboniferous rocks of the eastern United States. Most of the two thousand species now living are tropical, with only a few in the northern woodlands. Some have successfully accommodated themselves to man, and they live the year around in his dwellings where, as agile as mice and as nocturnal in their habits, they have been able to hold their own until the invention of modern insecticides. Since they are developing resistance to these, perhaps the issue is not yet decided.

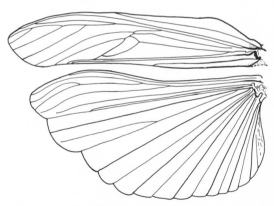

FIG. 35. *Wings of a grasshopper.*

The northerner has little idea of the diversity of these insects. Some of the tropical species are brilliantly colored. Some are small, a few millimeters in length, with narrow, fringed hind wings like those of the minute thrips or certain small moths. Others are among the largest insects; one (a species of *Blaberus*) from the West Indies and the Florida Keys exceeds two inches and scurries across the floor with an impressive rustling noise. This animal is shipped by airmail from suppliers in Florida to laboratories for use in research on various physiological problems. As a diverse group of primitive insects, the study of cockroaches probably would yield much interesting information relevant to the early history of winged insects, but they are comparatively neglected.

Although the integument of the cockroach is flexible, it is tough and difficult to cut even with the dissecting knife. The surface is oily and rather slippery, and many species secrete malodorous substances. Among the characteristic features are the very wide front part of the thorax, which shields the head, and the peculiar position of the head, which is reflexed so that the mouthparts are directed backward. The head, however, is mounted on a slender and mobile neck, so that the mouthparts can be brought forward into a normal feeding position. The ocelli are unusual among the Orthoptera in that there are only two, as in the Thysanura. They usually are degenerate in the many flightless cockroaches. The cerci are well developed; most of the experiments on the sensory function of cerci have been made with cockroaches.

Cockroaches enclose their eggs in double rows in an elongate capsule which is formed inside a pocket in the abdomen. Two large glands opening into this cavity each produce different kinds of liquids. When the two liquids mix as they are poured around the eggs, they combine chemically to produce a tough, hard protein. This is molded into the egg capsule by a device near the end of the abdomen shaped something like the jaws of pliers. Some species drop the capsule before the eggs hatch, others retain it so that the young are born alive.

Little is known of the biology of the great majority of the species, but some of the domestic roaches are well-known laboratory animals. They are convenient for nutritional studies: because of their omnivorous habits, they can be fed a variety of prepared diets in different physical states and, since they are small, can be fed costly, chemically defined diets that make it possible to determine nutritional requirements with precision.

Some of the North American forest-dwelling roaches live in small colonies. They feed on wood, which is digested by symbiotic protozoans that live in the hindgut. Since the protozoans are not present in the insect when it hatches, and are subsequently lost with each molt, community life is necessary for infection and reinfection with these essential microbes, which are picked up by eating fecal pellets. The wood-eating termites have similar protozoans.

About three fourths of the species of Orthoptera belong to the suborder Saltatoria, which have the femora of the hind legs enlarged to house the powerful leaping muscles. These hungry, abundant, and noisy animals are among the best-known insects. The shrilling and rasping insect calls that fill the air on summer nights are the songs of these orthopterans. Along the dry foothills of the western mountains one may walk for miles through the parched grass with a steady cascade of grasshoppers parting ahead, some rising in strong flight, others tumbling aside with clumsy leaps. In the great locust plagues they settle with a roar of wings to lay waste all vegetation.

There are three divisions of the suborder. The family Acrididae, that of the grasshoppers or locusts, is one of active, diurnal insects, that leap or fly strongly when threatened (Fig. 36). Their calls usually are muted and are not much noticed. Structurally they are marked by the short antennae, the location of the ears on the abdomen, and the short ovipositor, which is used to force entrance for the abdomen into the ground, where the eggs are placed. The two other families, the long-horn grasshoppers (Tettigoniidae) and crickets (Gryllidae), are mostly nocturnal, are relatively sluggish, and are loud and accomplished songsters. They have very long, slender antennae, ears on the front legs, and a long ovipositor used to insert eggs in the ground or, more often, in plant tissues.

In early spring one sees in meadows and pastures the bright-winged locusts that dart away in erratic flight, displaying wings of deep red or yellow. The grasshoppers that overwinter as nearly grown nymphs

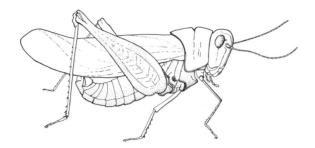

FIG. 36. *Acridid grasshopper.*

and appear early in the season as adults are in the minority, however, and it is not until late summer that the grasshoppers—perhaps twenty or thirty species in a single field—become the dominant insects. The partiality to warmth is shown also in their world-wide distribution, for most species are tropical. Although some live in the foliage of trees and shrubs, the acridids are characteristic of open grasslands, where their strong flight and ability to leap serve them well in the presence of a number of predators.

These defenses usually are supplemented by protective coloration, the front wings and rest of the exposed parts of the body closely matching the background. Even the brightly colored hind wings of some of the grasshoppers seem to give the end result of concealment. A bird pursuing a red-winged locust would be confused by the sudden disappearance of the color when the insect alights and folds the hind wings under the wing covers. The entomologist T. D. A. Cockerell writes: "In addition to this, they suddenly double back on alighting, so that they are not where we might expect them to be. I have been completely deceived by this; possibly birds are more intelligent."

The group of grasshoppers that usually have such wings—the subfamily Oedipodinae—live in sparse grasslands where there is much bare ground. They rest on the ground instead of in or on tangles of vegetation as do the other grasshoppers. When at rest, with the colored wings hidden, they are concealed by the remarkably close match between their color and that of the background. Even the eyes are colored like the rest of the animal. The observer is further impressed by the fact that individuals of a species living in adobe soils will be pale gray, whereas those of the same species on soils derived from the weathering of red rocks will be reddish. Laboratory experiments demonstrate that such grasshoppers can to a degree control their color while they are growing up. Those grown in cages with a dark background, for example, will turn out to have permanently dark integuments, those in light cages light ones.

The terms locust and grasshopper are more or less interchangeable, but, applied most strictly, locust refers to those species of the family Acrididae that appear in dense migratory swarms. Each continent has, or has had, one or more species of destructive locust. In Eurasia an important species is *Locusta migratoria*, which occurs from Europe east to the Philippines. In some years this migratory locust disappears, then, at irregular intervals, suddenly appears in great swarms. The mystery of its appearance was solved by the discovery that the grasshopper leads a Jekyll-Hyde existence. There is throughout Eurasia a harmless, solitary grasshopper given the name *Locusta danica*, a sedentary type with relatively short wings. It now is known that this type under certain conditions becomes changed into the long-winged, gregarious, restless *migratoria*. The two are thus phases of a single variable species. The change into the long-winged migrant takes place when favorable climatic conditions allow populations of the *danica* phase to build up to very high levels. When the young grasshoppers grow in crowded conditions, various influences, apparently visual stimuli and jostling by others, cause development to alter in such a way that the wings become longer, the pigmentation brighter, and the instinctive behavior becomes that of the migrant. The insect has, so to speak, two sets of hereditary materials, each capable of producing an insect adapted to one of two modes of life. One set of environmental conditions causes one array of genetic material to come into operation; another set switches development into the other channel. An important breeding ground where *danica* becomes transformed into *migratoria* is in western Asia, in the great reed beds on the shores of the Aral and Caspian Seas.

In North America the migratory locust was a smaller species,

Melanoplus spretus. Early pioneers in the West saw great shifting
black clouds of this insect descend on their new farmlands and destroy
whole fields of crops in a few hours. The government entomologist,
C. V. Riley, sent out to investigate the biology of this plague locust,
suggested as an emergency measure that the farmers, to avert the real
danger of starvation, turn the tables by eating the locusts and tried
several recipes on his friends. But the idea did not catch on.

Oddly enough, this plague locust, which was, like the passenger
pigeon, once incredibly abundant, is now apparently extinct and has
not been collected for more than fifty years. It was thought that the
locust might be the long-winged phase of the still-existing *Melanoplus
sanguinipes*. When crowded in cages in the laboratory, this grasshopper
produces individuals with long wings, like those of *spretus*. However,
careful examination of the structure of *spretus* and laboratory-grown
sanguinipes shows some important differences. Possibly the migratory
locust still exists in small colonies in the Rocky Mountains, but these
have not been discovered. Individuals can be collected frozen in the
ice of Grasshopper Glacier, Montana—or at least could be until
recently. Visitors to this glacier report it nearly gone, with the grass-
hopper-bearing face inaccessible on account of falling rocks.

Most grasshoppers are rather soft-voiced, the usual note being a
soft buzz made by rubbing a row of pegs on the hind legs against the
front wings. Others make a good deal of noise by scraping the wings
together while hovering high in the air. A characteristic sound of a
hot summer day in the dry mountain ranges of the West is the cheerful
crackling of these "firecracker" or "castanet" locusts.

The long-horned, or tettigoniid, grasshoppers live in vegetation or
on the ground and have loud and persistent songs. The many species
that live in foliage are usually green.

One of the best-known insects of the Eastern United States, although
more often heard than seen, is the katydid, whose raucous cries fill
the air on hot summer nights. These magnificent bright green insects
are most likely to be seen after the first frost, when they are stunned
and fall from the trees.

The songs of the long-horns are produced by the vibration of an
elastic membrane, the tympanum, which is part of the base of the

right front wing. When the bases of the wings are shuffled rapidly together, a file on the base of the left wing sets the tympanum vibrating. Females are without this apparatus. These nocturnal orthopteran singers can be traced down with a flashlight. They are quite wary, becoming silent at a careless step, but will strike up again if one waits quietly and will continue singing even when fixed in the beam of light.

The crickets, like the long-horns, will rarely be made to fly but try to escape by concealing themselves. They differ structurally in that the wings usually lie flat, rather than sloping and roof-like, and the ovipositor is cylindrical rather than sword-shaped. The familiar North American crickets, at least, have high-pitched, musical, chirping or tinkling songs, whereas the long-horns specialize in rasping or buzzing notes. The cricket songs also are made by rubbing together the bases of the front wing. With their love of the miniature, the Japanese keep these songsters in small cages and prize them for their cheerfully musical company. In Europe they are also sold as pets. The males are combative, and in the Orient fights are staged in the manner of cock fights. It is said that large sums have been wagered on battles between famous crickets.

The song functions in courtship and may also serve to establish territory, like the songs of birds. Its mating function is demonstrated by the fact that a female cricket may be enticed to a telephone with a male singing at the other end of the line. Different species of crickets have different songs that can be recognized by the expert. Recently developed electronic techniques make it possible to record these short, quick, complex songs graphically. There are minute morphological differences in the sound-making areas on the front wings of species that have different songs, but the differences between the electronic recordings of songs are much more evident and more easily analyzed.

Males of the species of *Oecanthus*, a group of pale yellow crickets that live in foliage, secrete a liquid (from glands that open on the top of the thorax) that is drunk by the female before mating. During courtship the male backs up to and under the female in such a way that the offering is conveniently available.

One of the most familiar crickets of the eastern United States is the black field cricket that sings both day and night and whose increasingly feeble chirps in late September remind the gardener of approaching frosts.

The one thousand species of the family Gryllidae, most of them tropical, are diverse, some not at all resembling the familiar house or field crickets. The mole crickets that live underground, burrowing with front legs that are highly specialized for digging, are so distinctive that they have been placed in a separate family. They are rarely seen, unless turned up by the plow, although the males may fly to lights in great numbers. Another group of crickets are minute, a few millimeters in length, and live in ant nests.

Most of the fifteen hundred species of praying mantids are tropical, but there are a few of the large and spectacular species in the North. The group is remarkably diverse, some with the body fashioned into leaf-like structures, some resembling flowers, others unadorned, inconspicuous ground dwellers. This purely carnivorous suborder of Orthoptera is characterized by the raptorial front legs, although there are some rare primitive species, somewhat cockroach-like in general appearance, that have the front legs little specialized.

The incongruity of appearance and deed, when the mantid devoutly chews its victims that are held in the supplicating front legs, apparently gives these insects the engaging quality that makes them popular pets. The male usually tries to approach the female undetected, to seize her unawares, but often he is seen, and the female then catches and eats him, usually beginning at the head. The loss of his head, however, galvanizes the male into action, and he can successfully complete copulation without it. This behavior pattern, in which she devours the male, is of obvious advantage to the female, and to the species, because she is able to put to good use an otherwise worthless mass of protein.

The remarkable leaf-camouflage of many of the praying mantids probably serves to protect them from birds or lizards rather than to conceal them from their insect prey, since the visual apparatus of the insect is not one that could appreciate such refinement. The bright color-display on the underside of the flattened body of certain mantids is, however, presumably directed toward other insects. These mantids take up a position among flowers and turn their body so that the color is visible. They even imitate the nodding of the flowers in the breeze by moving their body. Flower-visiting insects that are lured to them are captured.

When the eggs are laid, in large masses, they are covered with a frothy material that soon hardens into a spongy, impenetrable shield. A small parasitic wasp of the family Scelionidae gets around this defense by depositing its eggs in the mantid eggs during the short interval between laying and hardening of the egg case. The wasp manages to be on the scene at the crucial time by riding on the mantid until the eggs are laid. The young of the wasp develop inside the eggs of the host.

Mimicry of vegetation is almost universal among members of the suborder Phasmodea. Some of them resemble leaves so closely that the natives of certain tropical areas think that they are born as buds on a tree, grow into leaves, sprout legs, detach themselves from the parent plant and walk away. Others are attenuated and twig-like. Yet others are black and conspicuous and defend themselves by corrosive secretions.

It is rather astonishing to see that the vegetable motif is continued into the eggs, which are laid separately, each surrounded by a tough, ribbed or ornamented capsule that gives it the appearance of a seed. The eggs are dropped broadcast. In the forests of the warmer parts of the Mississippi basin these insects are so numerous that their eggs may be heard falling to the forest floor like rain.

Like some other insects, the phasmids are able to regenerate legs or antennae that have been lost. They also have the ability to cast off a leg (autotomy) along a preformed line of weakness, an adaptation for escape from predators. If an antenna is cut off, a leg grows in its place. The ability to regenerate appendages is lost when the animal becomes mature and ceases to molt, but, if the corpora allata (a hormone-producing gland) from a young insect is implanted, both molting and the ability to regenerate lost appendages are restored.

There are about two thousand species of this strictly herbivorous suborder known. The group is most diverse and abundant in the Australian region.

The peculiar grylloblattids, or ice crickets, which are not closely related to any other orthopterans, were discovered in the early 1900s in the mountains near Banff, Alberta, in Canada. Since then a few more species have been described from the mountains of the Far West and of Japan, where they have usually been found under stones at the edges of snow or ice fields. They are completely wingless and look remarkably like overgrown thysanurans. The legs are adapted for running.

From primitive Orthoptera-like insects several orders have arisen that retain the generalized mouthparts of the Orthoptera, their primarily terrestrial mode of life, and their primitive mode of development, in which the young are essentially like the adults. These orders are the Dermaptera, the Embioptera, and the Isoptera.

The earwigs (Dermaptera), given a bizarre appearance by the pair of strong pincers at the end of the abdomen (Fig. 37), are closely

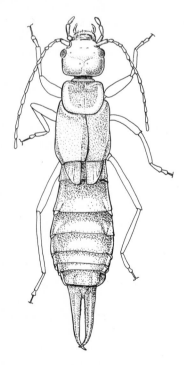

FIG. 37. *Earwig.*

related to the Orthoptera. The orthopteran wing covers are here modified to an extreme, the front wings being thick, hard, and without wing veins. Since the wing covers are very short, the long, semicircular hind wings have to be folded both lengthwise and transversely for storage. To a person who has difficulty folding road maps, the feat looks impossible.

The characteristic forceps are modified cerci; one peculiar family of earwigs found only in certain bat caves in the Orient still has ordinary jointed cerci instead of pincers. Although the forceps are probably important in the life of the earwig, there is some uncertainty about their function. When disturbed the earwig will attack with its forceps, but they make quite weak pincers; and it has been suggested that they are used to catch prey and to assist in folding and unfolding the wings.

The most aberrant of the earwigs are the hemimerids. These live as parasites on certain African rodents and eat hair or sloughed-off skin. They sometimes are put in a distinct order, the Diploglossata, which, with its two species, would have the distinction of being the smallest order of insects in existence. The cerci are not shaped into pincers. The young are born alive and are nourished before birth by a structure that functions like the placenta of mammals.

The typical earwigs are oviparous. They brood the eggs and also care for the young. Most of the approximately one thousand species are tropical. These nocturnal insects live under logs, stones, and other out-of-the-way places. In the western United States an introduced species has become common enough to be a garden and household pest.

Like the earwigs and a few of the cockroaches, the insects of the order Embioptera show glimmerings of the social organization that is so highly developed in termites. The embiopterans are not truly social insects because they lack caste differentiation, a matter that will be taken up when considering the termites.

Embioptera are small and soft-bodied. The hundred or more known species are confined to the tropics and warm temperate regions. The colonies are usually in concealed places but may be exposed, as on the trunk of a tree. Apparently one of the functions of the colonial habit is to pool resources in order to construct an elaborate dwelling, a several-layered network of silken tunnels. The function of this web is not fully understood, but it has been suggested that it protects the inhabitants from small insect predators not so skillful as themselves in traversing the webby maze. The silk is spun from glands in the peculiarly enlarged front legs.

When disturbed, the embiopteran may display its ability to run forward or backward with equal speed, a facility doubtless useful to elongate creatures living in narrow tunnels.

The males are winged, the females wingless. Like the termites, they have the front and hind wings alike, both membranous and equally important in flight (Fig. 38).

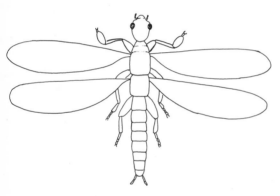

FIG. 38.   *Order Embioptera.*

Four major groups of insects—termites, ants, bees, and wasps—have separately evolved, complex societies in which individuals that do not take part in the sexual life of the species labor to build the communal shelter and to care for the young. The differences in function are reflected in differences in structure, so that in the truly social insects there are two or more recognizable castes. The most primitive of the social insects is the order Isoptera, or that of the termites. The primitiveness of the termites (commonly called "white ants") lies only in their structure, for their social organization is complex and highly evolved. Although there is no fossil record to prove the supposition—the oldest fossil termites known are of early Tertiary age—it is believed that the termites are considerably more ancient than the other social insects. They probably have been in existence for well over 100 million years.

There are several groups of mammals that are highly specialized for feeding on termites: the South American anteaters, including powerful ground-dwelling animals and small arboreal kinds that feed on tree-nesting termites; the South African aardvark; the armored pangolins of the Old World; an Australian marsupial; and the egg-laying spiny echidna of Australia and New Guinea. The great differences between some of these and other mammals indicate a long history for the termite-eaters. The Australian marsupial anteater is primitive even for a marsupial and has structures reminiscent of certain very primitive Mesozoic mammals.

Their close relationship to the cockroaches suggests, but does not prove, an ancient origin for the termites. The Protozoa living in the digestive tract of the more primitive, wood-eating termites are like those of the wood-eating cockroaches. Although modern termites characteristically have the hind wings narrow, without the wide, folded area of the wing of cockroaches and other Orthoptera, there are fossil termites with roach-like hind wings, and the most primitive living termite, *Mastotermes darwiniensis*, has such wings. An Australian burrowing cockroach sheds its wings when mature, as do the termites.

The order Isoptera is a small one from the standpoint of number of species (about fifteen hundred), but the termites are extraordinarily abundant in the tropics, where their dwellings may dominate the landscape and where hardly a square yard of earth can be dug into without turning up some of these insects. They are an important force in converting the huge amounts of wood produced in the tropics back into the components of the soil and air.

Termites differ from the other social insects in that the nonreproductive castes are imperfectly developed individuals of both sexes. In the ants, bees, and wasps, only imperfectly developed females form the worker and soldier castes. In the most primitive of the termite groups, the workers are merely immature individuals, which go on to complete their development. In the more advanced termites, however, growth of the workers is permanently arrested, so that the workers represent a true caste.

The workers of a colony, then, are typically males and females that are unable to reproduce. They are wingless and have mandibles of the chewing type, which they use in gathering food and in constructing the nest. In most species the workers are eyeless and never come out in the sunlight. They tunnel from their nests to supplies of food or, when forced to traverse exposed surfaces, build a covered passageway. A soldier in New Guinea took his gas mask down from a peg on the wall of a hut and was greeted with a swarm of termites that poured out of the mask; only then did he notice the neatly made covered

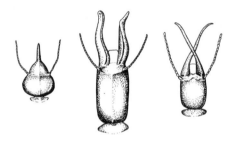

FIG. 39. *Heads of soldiers of various species of termites. The head at the left is that of a nasute, the other two of mandibulate soldiers.*

passageway that ran from the floor up to the mask. Because they are thus protected from drying, such termites can get by with a soft, relatively permeable cuticle. A soft cuticle is usually light colored; hence the name "white ant." Some of the primitive termites forage in the open in the manner of true ants; these have the normal hard and dark cuticle and have eyes.

As the name implies, the soldier caste is believed to have the function of defending the colony. One kind of soldier, called the nasute, has the head drawn out into a nozzle through which it squirts a sticky repellent substance (Fig. 39). The gland that serves the nozzle may occupy most of the space inside the body of the soldier. Another type of soldier has large mandibles, which may look like formidable weapons, or may be curiously and incomprehensibly twisted in such a way that one can not imagine a function for them.

At intervals a colony produces enormous numbers of sexually mature, fully winged males and females (Fig. 40). Perhaps governed in some way by weather conditions, the winged termites of a given species pour out of their nests at the same time. The incredible numbers of these weakly flying, succulent insects draws many predators to the scene. One observer describes a bird sitting helplessly stuffed amid such a swarm, with termite wings protruding from its mouth. However, there need be only a few survivors from each colony to perpetuate the species, and the slaughter is the price paid for dispersal to new grounds and the fresh variability that results from mixing heredities in sexual reproduction.

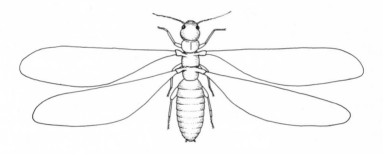

FIG. 40. *Winged termite.*

When the flying termites come to earth, they break off their wings near the base, along preformed lines of weakness. After one of these synchronized dispersal flights of an abundant species, the ground may glisten for miles around with discarded wings. After losing the wings, the termites pair off, then search out a shelter that will become the nucleus of a new colony. It is said that the pair, after building the nest chamber, chew off the tips of their antennae. Perhaps the significance of such a ritual would be that, added to the self-inflicted mutilation of breaking off the wings, it commits them even more firmly to an underground, sedentary life.

Unlike the queen bee, which usually mates only once and which fertilizes the eggs laid during her life from sperm stored in the spermatheca, the termite queen mates frequently with her original consort.

After fertilization, the queen increases in size as her ovaries swell. In some of the more advanced termites, the queen grows from an original size of a fraction of an inch to a helpless, swollen egg factory three inches long that pours out eggs at the rate of several thousand a day for many years.

The first young are fed by the royal pair on predigested wood or on materials stored in the fat body or derived from the degenerating wing muscles. These become the first workers, and the parents gradually relinquish the work of the colony to them.

Although the factors that will determine the caste of the young termite born into its society are not yet well understood, it is believed that they are extrinsic, that they exert their influence after the egg is laid. In other words, caste is not hereditary. It is believed that food and, in some way, the influence of the other termites of the community, determine whether the young will become soldier, worker, or a winged reproductive. Experiments with certain primitive termites give such results as these: thirty or forty young termites taken at random from the colony will produce a king, a queen, and an assortment of workers and soldiers; if two young termites, one of each sex, are isolated, one will become a king, the other a queen; if both are of the same sex, only one will become sexually mature, and the other becomes a worker.

The habitations of termites are of several kinds and, like the webs of spiders, vary according to the species, so are of some significance in classifying them. The most simple are excavations in wood or in the ground. Aboveground termitaria are sometimes imposing structures—one in Australia was twenty feet high and twelve feet in diameter. These are constructed of earth particles cemented together by a salivary or anal secretion into a concrete that can hardly be broken with a pick. Some built in trees are of bits of wood glued together.

Wood is the main food of termites and is, as has already been described, digested by the huge mass of protozoans living in the hindgut of the insect. Seeds and bits of leaves also are gathered by some termites. One species forages at night for lichens. The most highly evolved termites, which lack wood-digesting protozoans, eat fungi that are grown in the colony, cultivated on finely chopped plant material.

15

A specialized sideline

The group of orders to be considered now are primitive in that they develop gradually, without strong metamorphosis, that is, the young, except for wings, is like the adult and is not specialized for a mode of life different from that of the adult. These insects differ from the Orthoptera and other primitive insects by having the mouthparts specialized for a liquid diet; this specialization is not evident in the most primitive orders of the group, which more or less bridge the gap between the Orthoptera and Hemiptera, but the Hemiptera, or true bugs, the largest order of the group, have the mouthparts so strongly modified that the original three pairs of feeding appendages are hardly recognizable.

Most reminiscent of the orthopteran line are the Zoraptera (Fig. 41), which resemble termites, a group closely related to the Orthoptera. These minute insects, about three millimeters long, live in colonies in rotten wood, under bark, or in the ground. They are either wingless or winged, and the wings are shed like those of the termites, leaving stubs on the thorax. It seems to be characteristic of them to live in the vicinity of a termite colony. The rare and few species (about twenty) are tropical or semitropical. There are two species in the southern United States. They retain the chewing type of mouthparts and, judging from the gut contents, feed on other insects, probably as scavengers. So far as known the zorapterans, although living in small colonies of twenty to a hundred individuals, are not truly social, since the existence of a sterile caste has not been demonstrated, but their biology is too little known to be certain of this.

Somewhat similar to the zorapterans, but more abundant and better known, are the psocids, or book lice, of the order Corrodentia or Psocoptera (Fig. 42). With its thousand or so species, it nevertheless ranks as a minor order, and, except for a few common species, its members are little noticed by any but the entomologist. Although it had from its structure been judged to be a rather primitive group, the fossil record gave no support to this idea until the discovery of the remarkable insect-bearing fossil beds at Elmo, Kansas, in the early twentieth century. In small pockets in the Permian rocks at this locality thousands of insects have been found, among them a varied lot of poscids.

These insects are jawed, like the termites, are soft-bodied, and are either winged or wingless, sometimes individuals of even the same species differing in this way. A peculiarity of the feeding apparatus is a pair of hard, slender picks, apparently used to gouge out pieces of food, that cannot be matched with the normal three pairs of feeding appendages. The correlation between wings and ocelli is well illustrated in this order, for the wingless species are invariably without ocelli, and the winged species have them. In the Orthoptera the

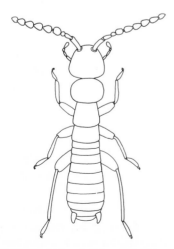

FIG. 41. *Order Zoraptera.*

young are equipped with ocelli, but in the Corrodentia the ocelli do not appear until the last molt, when the insect becomes mature. In some species the matter of suppressing or retaining the ocelli is in an unsettled state. One with rudimentary wings, that normally lacked ocelli, in one locality had over 10 percent of the individuals equipped with ocelli. The suppression of ocelli in the young is characteristic also of the Hemiptera, and none of the larvae of the insects with complete metamorphosis have true ocelli.

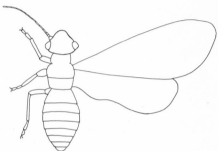

FIG. 42. *Order Psocoptera.*

Most species are only a few millimeters long, but a South American group has a wingspan of an inch. They typically live in out-of-the-way places, eating fungi or scavenging dead plant or animal material.

The Anoplura are small, wingless external parasites of warm-blood animals, so specialized for the conditions of life found on their host that they can live away from it for no more than a few days. The pigeon louse (*Columbicola columbiae*), for example, thrives at 37°C., but the young die in a few days if the temperature is lowered to 33°C. The order Anoplura is divided into two suborders (by some writers considered to be orders), which correspond to two different modes of life.

Nearly all of the nearly three thousand species of biting lice (suborder Mallophaga) are parasites of birds (Fig. 43), but some live on mammals. With their jawed feeding appendages they eat hair, feathers, the dead outer layers of the skin, and dried blood. A freshly killed bird may swarm with these minute insects, which run with agility between the feathers. As the body of the host cools some leave it, only to die within hours or days.

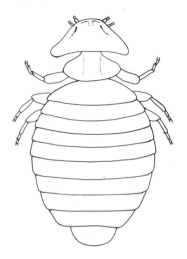

FIG. 43. *Biting louse of the suborder Mallophaga, from a bird.*

Usually there is little likelihood that the lice of one species of bird will transfer to another distantly related species. Therefore, the classification of the biting lice to some extent parallels that of the birds, so that a family or order of birds may have one or more groups of lice peculiar to it. The systematics of these insects is therefore of interest to the ornithologist, since it provides some clues for classifying birds. On a single bird, different species of lice may be specialized for living in different regions of the body, as the neck feathers, the region under the wings, or the down on the breast.

The mallophagans that live on birds have paired claws of the normal insect type. The relatively few species of biting lice that live on mammals generally have lost one of the claws and use the single claw to grasp hairs as they clamber through the fur of their host. The sucking lice, of the next suborder, also have this adaptation. It is interesting that the primitive marsupials of Australia have biting lice equipped with the primitive paired claws.

The five hundred species of sucking lice (suborder Siphunculata) are all parasites of mammals (Fig. 44). Even the water-going seals are infested with them. Unlike the agile mallophagans, they are sluggish insects.

The species of louse that lives on man (*Pediculus humanus*, or

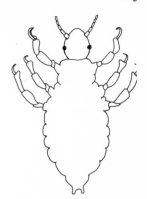

FIG. 44. *Louse of the suborder Siphunculata, from a mammal.*

"mechanized dandruff") is subject to a disease caused by one of the rickettsias, and contracts the disease by drinking the blood of a man infected with this microorganism. Should the louse survive this often fatal disease, it carries the microbes in latent form in the digestive tract. When a man harboring lice scratches louse feces into skin abrasions or gets the material in contact with moist eye membranes, he may himself become infected with the rickettsia, which causes the very serious disease called typhus.

A peculiar group of lice that live only on elephants are intermediate in structure between the biting and sucking lice and are sometimes placed in a suborder of their own.

Although the Thysanoptera, or thrips, are usually considered to be closely related to the Hemiptera, they are one of the most peculiar orders of insects, with aberrations of structure that may indicate quite different relationships. Fossils have been found in Permian and Jurassic rocks of Russia and Turkestan, but these extinct thrips apparently do not throw any light on the origin of the group.

Thrips usually are very small insects, from one to a few millimeters long, although there are comparatively gigantic Australian species that reach half an inch. When a thrips is disturbed, it may curl up the tip of the abdomen and wave it about in a threatening way reminiscent of the rove beetles, but apparently one of the functions of this behavior is to use the combs at the sides of the abdomen to arrange the long hairs on the wings for flight. The wings are narrow straps,

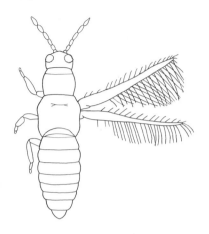

FIG. 45. *Thrips. Order Thysanoptera.* (*After Bailey.*)

but the fringe of hairs greatly increases the wing surface (Fig. 45). A set of hooks on the base of the hind wings couples the two pairs in flight.

The compound eyes resemble a cluster of grapes, the individual facets being round and bulging, without the hexagonal outline of the closely crowded ommatidia of other insects. The mouthparts are carried pressed against the underside of the thorax. They are the most Hemiptera-like feature of the Thysanoptera, for the mandibles and parts of the maxillae take the form of slender, piercing stylets. Usually the right mandible is degenerate, so that the mouthparts are asymmetrical. There is a fundamental difference between the feeding mechanism of the Thysanoptera and Hemiptera, however, because the stylets of the thrips do not fit together to form a drinking tube. The channel through which fluids are drawn is in the thrips formed by the labrum and labium, while in the Hemiptera it is formed by the opposed maxillary stylets. A piercing beak has originated independently more than once in the history of insects; not only does such a structure appear in the Thysanoptera, Hemiptera, Anoplura, and Diptera, but also in the extinct Protohemiptera of the Paleozoic and Mesozoic. The name Protohemiptera is inappropriate, for it is believed that these insects were not at all related to the Hemiptera.

The best-known of the two to three thousand species of thrips are vegetarians, feeding on plant juices and perhaps pollen. Many kinds live concealed in flowers, especially flowers with tubular corollas. In the eastern United States one of the plant-eating species becomes extremely abundant about wheat-harvest time. It is so small that it is scarcely deterred by window screens and swarms into houses. One entomologist granted an interview to the local newspaper at the time of such an outbreak and gave them a good deal of scientific information regarding the insect, to be rewarded with a headline that read: "Scientist says they don't bite, but we know better." Logically, these vegetarians should not bite, but they do, having a peppery bite all out of proportion to the size of the insect. There also are many predaceous species that feed on mites or small insects and their eggs.

It is in their metamorphosis that the thrips differ most from the Hemiptera, and from all other orders so far discussed, because the young are without external wing pads, and there is a quiescent, non-feeding pupal stage. Usually their metamorphosis is considered to be of an intermediate type, because before the pupal stage there is a stage with external wing pads, called the prepupa, in which the insect may be more or less active, but in which it does not feed. However, the prepupal stage is characteristic also of some insects with complete metamorphosis, such as the bees, where it is the stage in which the wings first become visible. The fact that the pupa of the thrips can be roused to activity, if disturbed, also is held to show that their metamorphosis is intermediate. Often the pupa of the thrips is sheltered in the soil, where it may be further protected by a cocoon.

The large order Hemiptera—the true bugs—with well over fifty thousand species, is a very diverse group but is given unity by the specialized feeding structures, which nearly always are adapted for piercing and for drinking liquids. Many species are able to eat such solid food as seeds, but they accomplish this by predigesting it, then drinking the dissolved nutrient. Both the mandibles and maxillae have been modified into stylets. The four fit together to form a piercing and sucking tube so slender that it is almost invisible in even a large bug. This hair-like beak is hidden, when not in use, in the head and in the stout, deeply grooved labium, which forms a supporting sheath for it. Sometimes the beak is so long that it is carried in a coil in the body cavity. The mandibles form the outer layer of the beak and may have locking devices at the edges to hold them into a permanent tube. The sharp, serrated tips of the mandibles assist in piercing. The maxillae held inside also may be locked together. On their inner faces are two parallel grooves that form two channels when the maxillae are opposed. Food is sucked in through the front channel, and digestive fluids are pumped out through the posterior channel.

There are two subdivisions of the order, sometimes ranked as orders: the suborder Homoptera, with the front wings not modified, and the suborder Heteroptera, with the basal half of the front wings hard and thickened, so that the wing veins there are almost obliterated (Fig. 46).

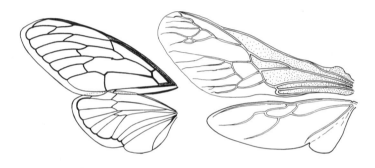

FIG. 46. *Wings of (left) a cicada (Homoptera) and (right) an ambush bug (Heteroptera).*

All of the Homoptera feed on the living tissues of plants, in which they imbed their very long beaks to feed in a leisurely manner. The suborder is widespread but in the tropics is most diverse and is there represented by many large and bizarre species. The group is probably more ancient than the Heteroptera. In the Upper Permian rocks of Australia they are the most abundant group of insect fossils and in the Triassic of a locality in Queensland are second in importance to the beetles. The oldest known Heteroptera are of Triassic age.

Such relatively primitive homopterans as the cicadas (family Cicadidae) (Fig. 47), spittle bugs (Cercopidae), leaf hoppers (Cicadellidae), and tree hoppers (Membracidae) are active insects able to fly strongly. Members of the last three families also are able to jump, and they spring away so quickly that they seem to disappear before one's eyes. Some are relatively large, especially in the family Cicadidae, where there are species three inches long. Such advanced groups as the plant lice (Aphididae) and scale insects (Coccidae) are very small, usually one to three millimeters long, and are sedentary, although in dispersal the weakly flying aphids may drift long distances in the wind.

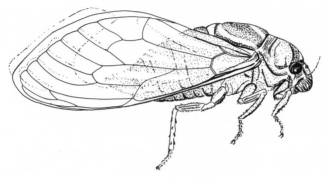

FIG. 47. *Cicada.*

The seventeen-year "locusts," or periodical cicadas, have appeared in incredible numbers in the virgin hardwood forests of the eastern United States. Indians told the early colonists about the regular timing of these plagues, which were to them portents of evil, so that, when Linnaeus described the species in the middle of the eighteenth century, he knew enough about it to name the species *septendecim.* Although the species in the North nearly always requires exactly seventeen years to complete its development, various local races have gone out of synchronization with one another, so that they appear in different years. Nearly every year one or another race of the periodical cicada appears somewhere in the eastern United States, and usually a given locality has more than one race, each separated from the others by a gap in timing rather than by spatial isolation. Since records have been kept for many years, entomologists can predict when the cicada will appear. Another periodical cicada, especially prevalent in the southern states, has a cycle of thirteen years.

The female cicada lays her eggs in small incisions she cuts in the twigs of trees. When the eggs hatch, the young drop to the ground and burrow through the soil to the roots of the tree where they feed. Here they live, with little wandering, until the spring of their seventeenth year, when they move up to near the ground surface. Here they wait for a few weeks, then, almost simultaneously (most of them within a period of a few hours) emerge from the ground, and climb a short distance up on the tree trunks. The adult climbs out of the tough nymphal skin and flies up into the trees. Here the males begin their singing, making a sound something like that of variable-speed circular saw, by vibrating a pair of membranes on the underside of the abdomen. The cicadas mate and die during the same summer in which they left the ground.

Even in the present remnants of the hardwood forests the periodical cicadas are extraordinarily abundant. The emergence holes under a single tree indicated that forty thousand cicadas had been living on its roots. This perhaps gives a clue to the reason for the remarkably long lives of these insects. It is only because they feed and grow with extreme slowness that a tree could support such a large population. Were this number of large insects (about one-and-a-half inches long) to feed on the tree at such a rate as to mature in a single year, they doubtless would kill it.

The advantage of having all individuals of the species coming out at once, and in enormous numbers, seems to be twofold: it is relatively easy to find mates, and their abundance defies the capacity of predators to destroy them. To attain these incredible population densities, the periodical cicadas have sacrificed rapid growth.

The evolutionary interest of the simultaneous emergence of these hordes of insects after long years of life underground is heightened by the recently discovered fact that each emergence is actually composed of three species, marked by differences in size, color pattern, and song. That is, what used to be called *the* seventeen-year cicada is three species, and the thirteen-year cicada also is a compound of three species. Just as there is selective advantage for all individuals of one species to come out at the same time—those that emerge in the wrong year have fewer or no mates and are more strongly exposed to predation—there also is selective advantage for the three different species to be synchronized with each other. First is the matter of saturating the predators, a version of the safety-in-numbers principle. Second, it has been suggested that the calling song of the male has evolved to the point that it is effective only in choruses of large numbers of cicadas, so that again large populations, even if of mixed species, afford optimum conditions.

In the United States there are many species of cicadas other than the periodical cicada, all with shorter life cycles of one to a few years. The seventeen-year cicada can be distinguished from other large cicadas by the uniformly dark thorax (no light markings above) and the red front margin of the wings.

Both the Cercopidae (spittle bugs) and Cicadellidae (leaf hoppers) are important insects of the grasslands. In a study of a twenty-five-acre pasture in New York State, it was estimated that the leaf hoppers got more nutrient from the grass than did the eight head of cattle that spent seven or eight hours a day in the field. In a grassland area of southern Michigan, the common meadow spittle bug, *Philaenus spumarius*, was by weight the leading insect. The young of the spittle bug live hidden in frothy masses produced by mixing air with a secretion that is poured from the digestive tract. Presumably this material affords protection from predators and parasites. So completely are the young spittle bugs adapted to life in this shelter that they die if removed from it, their thin skin being unable to prevent water loss.

The small, feeble, soft-bodied plant lice, or aphids, are one of the most successful groups of Homoptera, with about two thousand species, many of which are very abundant. They have three outstanding characteristics that go far toward explaining their success: they have a high reproductive rate, relying mostly on parthenogenetic reproduction; they have complex life cycles by means of which they can be active during the entire growing season, the different forms migrating from one species of plant to another; and they have established a symbiotic relationship with ants in which these insects protect them from predators and care for them in other ways.

In a large mass of aphids crowded on the stem of a plant one does not have to wait long to see the birth of an aphid. It is born a miniature plant louse, ready to begin feeding. It has already begun to reproduce, for even before birth there are embryos in its oviducts that have begun development. Generation after generation of females, with perhaps ten days between each, are produced in this way, without the intervention of males.

The sometimes complicated life cycle of aphids is based upon polymorphism. The number of forms that a single species may take during its life cycle may be as high as twenty. More than one form may exist at the same time—one living on leaves, another on roots, for example—or they may follow one another. A common type is one with relatively few forms and is one in which the aphid is able to utilize two host plants during the growing season. The life cycle of such an aphid begins in the spring with a wingless female that hatches from an egg laid the previous fall. Without mating, she gives birth to perhaps fifty to a hundred young. These also are wingless and female and again without mating give rise to a generation like themselves. Generations of virgin and prolific wingless aphids follow quickly through the spring and early summer. The young, first a few, then all, now begin to develop wings, and soon the host plant is deserted as they scatter to find the other species of plant on which the life cycle can be completed. On finding and becoming established on the new host, the winged females produce wingless females, and the generations proceed as in the spring.

With the onset of cold weather, the developmental mechanism that produces wings is again brought into action. The winged females reverse the flight of midsummer and seek out the first host plant. Here, for the first time, their young include both males and females, the males small and winged, the females wingless. These mate, and the female lays a few eggs, or sometimes a single large egg, that overwinter.

The appearance of wingless or winged generations of such aphids may be controlled in the laboratory by altering the proportion of light and darkness, or photoperiod. When the aphids are exposed to light corresponding to the length of day of the time of year they normally migrate, winged individuals are produced. Temperature alters the expression of this reaction to light, and low temperatures are important in bringing about the appearance of males in the last stage of the life cycle. In the production of female offspring, each individual receives two sex chromosomes. The physiology of the aphid is altered in such a way by environmental conditions of early fall that the sex chromosomes tend to separate imperfectly when the chromosomes are distributed during the cell divisions that produce the eggs. Among the types of chromosome distribution that result is one in which one of a pair of egg cells (actually the small, nonfunctional polar body) receives three sex chromosomes, the other one sex chromosome. The egg cell that has the single sex chromosome develops into a male. The sperm cells produced by the male are of two types, one without the sex chromosome and capable of producing a male when it fertilizes an egg, the other with the sex chromosome and therefore female-producing. However, all cells of the first type die, so that the fertilized overwintering eggs give rise to females.

By migrating from one host to another, the aphids not only get the advantage of leaving a plant that is past its prime but also leave behind the predators that have become established in the aphid colony. Larvae of ladybird beetles, of chrysopids, and of syrphid flies are among the most important of these predators. As these larvae grow, their voracity tends to outstrip the rate of increase of the aphids, which escape extinction by migrating. On the deserted site of the colony numerous brown corpses of aphids are left behind. Eventually a circular trap door is cut in the body wall of these, and a minute parasitic wasp emerges. Such parasites, members of the family Braconidae, are important natural enemies of aphids.

When the aphid feeds, it drinks large quantities of fluids and eliminates a liquid rich in sugars, called honeydew. If a tree is heavily infested with aphids, they varnish the leaves, and automobiles under the trees become covered with the sticky substance, usually mistaken for an exudate of the tree. Bees lick up honeydew to store in the hive; to the human taste the resulting honey is inferior. A similar sugary secretion of the related scale insects accumulates under bushes in the desert regions of the eastern Mediterranean countries, where it is mined for sugar, and is said to be the Biblical manna.

Many ants rely on aphids for food; sometimes they eat the aphids for meat but more often treat them like milk cows. The ant taps the aphid on the back with its antenna, and the aphid responds by exuding from the anus a clear drop of honeydew, which the ant drinks. Ants attending aphids are unusually aggressive toward outsiders: the ant that only tries to escape when on a normal foraging expedition will bite if one disturbs it at the aphid pasture.

Some species of aphids and ants are almost completely dependent on one another. The common cornfield ant, *Lasius americanus*, is an important pest of corn because it feeds on that plant indirectly by way of the corn root aphid, *Anuraphis maidi-radicis*. This aphid lives on the roots of various grasses and weeds as well as on maize. Generation after generation of wingless females live on the roots during the summer, then in late summer winged females crawl out of the ground for a dispersal flight. Wingless males and females appear in the fall, and the shiny, dark green eggs, which carry the life cycle through the winter, are gathered by the ants. During the winter the ants care for the eggs judiciously, carrying them from place to place as moisture and temperature conditions change. When the eggs hatch in the spring, the ants carry the young aphids to the roots of their food plants. During the summer the ants distribute the aphids more widely as the herd increases, and it is believed that this is primarily responsible for the underground spread of the aphid in corn fields. Aphids above ground appear to be helpless, but an ant that finds one carries it in its jaws to a feeding station. Probably the ant relies on honeydew for most of its food. The best way to control the aphid is to kill the ant, and this may be accomplished with deep plowing and thorough disking in the spring to break up the ant nests.

One of the aphids drastically affected the economy of France in the nineteenth century when it nearly destroyed the vineyards. This was the grape phylloxera (*Dactylosphaera vitifolii*), an aphid with a two-year life cycle: one year it lives on the roots—it is this phase that kills the grape vine—and the next it infests the leaves. Another difference between the phylloxeran and other aphids is that the parthenogenetic female lays eggs rather than living young; for this reason, it is sometimes put in a separate family Phylloxeridae. The

French government appealed to the public and offered large prizes for a remedy for this pest, only to get a flood of suggestions that took more effort to sort and evaluate than they were worth. The problem was solved by first properly identifying the insect, which was found to be identical with a North American species that lived on wild grape vines. The roots of the North American grapes are resistant to the phylloxeran, and, when compound grape plants were made by grafting the desirable European varieties onto the American roots, the growers brought the phylloxera problem under control.

The tendency among Homoptera, exhibited by the aphids, toward a sedentary life goes to an extreme in some of the scale insects (family Coccidae) where, after a brief period of wandering, the young insect drives its long beak deeply and permanently into the host plant, then loses its legs and antennae in subsequent molts. It covers itself with a shield of wax or proteinaceous material, so that it has little resemblance to an animal. There is, however, a continuous spectrum of types connecting this specialized condition with one in which the adult retains the antennae and legs, is mobile, and is not covered with a scale.

Parthenogenetic reproduction is common in scale insects, to the extent that in many the male is rare or unknown. The males are two-winged, with the hind wings reduced to a pair of short straps.

The scale may be a protective case for the eggs, which are laid in the end of the chamber. In others, the body of the female itself serves as an egg case, with the young hatching after the death of the parent. The scale is made of cast nymphal skins cemented together with wax and proteins or of filaments and grains of wax.

The scales of some species are important in industry. In China and India candles are made from waxy scales. Most of the Chinese wax, nearly three thousand tons annually, is produced in the mountains of Szechuan province. Here the scale insect (*Ericerus pe-la*) can not complete its life cycle and dies out in the fall. The following spring runners bring in by night a fresh supply of the young from Kiangsi, nearer the coast. Industry uses many tons annually of shellac, taken from the scales of another Oriental species. The bodies of a scale insect of Mexico that lives on cactus are deep red, yielding the dye cochineal.

After this dye was discovered by the Spaniards, the insects were brought to Europe for cultivation, where they supplanted another scale insect, *Kermes*, which had long been used for its red pigment. The invention of synthetic coal-tar dyes nearly destroyed the cochineal industry, but the natural dye still finds use as a harmless food coloring.

Insects of the suborder Heteroptera feed on both plants and animals. The most specialized predators capture other insects with their raptorial front legs; others suck the juices of sedentary larvae, pupae, or eggs as they are encountered. Yet others drink the blood of large vertebrate animals.

Most Heteroptera defend themselves with stink glands. In the young, these glands open on top of the abdomen; in the adult, with wings covering the abdomen, they open on the underside of the thorax. The area round the opening of the thoracic glands is often curiously ornate, with folds and roughenings. This is the evaporating surface for the odorous and volatile defensive secretion.

Several families of bugs live in the water, and, except for a brief dispersal flight, they are aquatic during the whole life cycle. All are marked by a distinctive morphological feature: the antennae are very short and in the water are carried concealed in pouches under the head. Probably the antennae are used only when the insect flies. The aquatic bugs get their oxygen supply at the surface, none being equipped with gills. Except for the aberrant corixids, or water boatmen, all are carnivores.

Water boatmen (Corixidae) feed mostly on microorganisms, which they draw up from ooze by movements of the very short mouthparts. They thus differ in feeding habits from all other Hemiptera. They swim well with long, oar-like middle and hind legs. The front legs, which are in the other aquatic bugs used to catch prey, here are used to make a chirping sound, probably in connection with mating. The eggs of these bugs are gathered in large quantities from the high lakes around Mexico City to be used for food.

The bugs of the family Notonectidae resemble the water boatmen but always swim upside down, powerfully propelled by the long hind legs. The color pattern of the back-swimmers is accordingly reversed: white on the dorsum, dark on the uppermost ventral side. These carnivores inflict a painful puncture wound if accidentally pressed against the skin.

The largest aquatic bugs are the Belostomatidae, or giant water bugs. One of these, the "fish-killer" of the Orient, reaches four inches and is sold dried for food. Other species, some nearly as large, are common in ponds in North America. With their well-muscled grasping front legs, these active bugs catch insects, tadpoles, and small fish. The female of some species lays her eggs in a compact layer on the back of the male, who carries them about until they hatch. In the water, the bugs lock the front wings in place over the hind wings, streamlining the body for underwater travel. In flight the wings are coupled by a clasp arrangement situated well out from the base of the wing. These very competent nocturnal fliers get far from water and are attracted to lights, which gives them the name electric light bug.

Water scorpions (Nepidae) are sedentary water bugs that wait for their prey. One of the best known is *Ranatra*, a common insect of ponds, which is built like an elongated praying mantid. When picked up, it holds out its long legs rigidly, in a kind of catalepsy, but the front legs are sometimes moved slightly to make a faint squeaking noise. This sound probably has some defensive significance; a similar noise is made, upon capture, by many thousands of species of insects of widely different groups. *Ranatra* usually clings to vegetation, with the long respiratory tube opening at the water surface, but it can descend and breathe for a long time from the air stored in the tracheal system. One specimen kept in an aquarium regularly ate flies, ants, and caddis larvae thrown in the water. When the insect came near, the water scorpion lashed out unerringly with a single front leg and seized its prey between the blade-like tibia and the femur. It would sometimes prudently catch another insect while feeding on the first and would drink an hour or more from each victim. The water scorpions, like some other aquatic bugs, have orienting devices that depend on differences in water pressure. These devices are fixed bubbles in longitudinal air channels that are widely spaced and have movable sensory hairs on the interface between air and water. Different water pressures, when the insect is tilted, compress the bubbles in varying degrees and move the hairs accordingly. The water scorpion *Nepa*, when on an underwater seesaw, keeps turning around so as to walk upward but does not do this when the orienting organs are destroyed.

The surface of the water is an extensive environment, rich in animal food. Of the several groups of insects that have become adapted to life in this environment, the best known is the family Gerridae, or

FIG. 48.   *An ambush bug, of the family Phymatidae.*

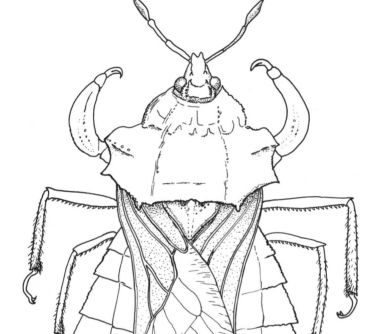

water striders. They search for floating insects or even leap into the air to capture low-flying prey. Unlike terrestrial insects, they use only four legs for locomotion. Since these all stay on the water while the animal skates along, it is always firmly supported. The weight of the insect causes the surface film to sag under the legs. The four dimples in the water cast large shadows on the bottom; one observer not yet acquainted with these bugs thought they had big feet that kept them on top of the water. If one adds detergent to a troop of water striders, they flounder as they break through the weakened surface film. Although most gerrids live on inland waters, the species of *Halobates* are at home on the open ocean. It is thought that they weather rough seas by diving below the surface.

Some of the Heteroptera inflict heavy damage on crops, so that the whole group has a reputation of being primarily herbivorous, but there are predatory species scattered through most of the families, and whole families are entirely predaceous. One of these is the family Phymatidae (Fig. 48), whose heavily armored members wait on flowers for their prey and can even overcome honeybees. The largest and most varied group of predators is the family Reduviidae, that of the assassin bugs. Its two or three thousand species are remarkably varied in color and form, but all bear raptorial front legs. When one of these bugs is picked up, it squeaks by rubbing the tip of the beak in a finely ribbed groove on the underside of the thorax. Most species prey on other insects, and their abundance and voraciousness make them an important component of natural communities. Several feed only on warm-blooded vertebrates. One of these, *Triatoma*, is well known because it is a vector for Chagas' disease, which is prevalent in Central and South America. Another, the South American *Rhodnius*, has achieved fame because it was used to demonstrate some of the basic facts of insect endocrinology. More specialized than these two for a parasitic existence are the bedbugs (Cimicidae), which are flattened like ticks and are wingless.

A few sweeps of the collecting net in a meadow will certainly yield specimens of the family Miridae, which is the dominant group of herbivorous bugs, with some five thousand species. Several of the species are nuisances in the garden and orchard, where they spot leaves with their feeding punctures and deform or kill fruit. Some of the most important agricultural pests are, however, in other families, as the chinch bug (Lygaeidae) and the squash bug (Coreidae). Probably the best known hemipteran is the box-elder bug, a red and black coreid that leaves its host plant, the female box elder, in the fall to swarm into houses in search of winter quarters. Chemical defense reaches its peak in the stink bugs (Pentatomidae); the species commonly seen are large hexagonal green bugs.

16

Ancient aquatics

Very early in the history of winged insects, the young of certain groups became specialized for a mode of life different from that of the adults: they invaded the fresh waters, leaving the adult stage to its aerial existence. The ancient groups with aquatic young and aerial adults are the orders Plecoptera (stoneflies), Ephemerida (mayflies), and Odonata (dragonflies). Such late comers to the aquatic environment as the Hemiptera, Coleoptera, and Diptera often breathe at the surface, but the early aquatic types taken up here have tracheal gills and are independent of the surface.

The stoneflies, mayflies, and dragonflies are undeniably primitive. They have a fossil record that extends back to Permian times and are closely related to several extinct orders that were on the scene at the beginning of the history of winged insects. For this reason they, like the Orthoptera, have always attracted students interested in the evolution of wings, of the wing veins, and insect flight.

Stoneflies (Plecoptera) live in pleasant surroundings; one associates them with trout streams and the pebbly, wave-washed shores of clear lakes. Unlike their near relatives, the Orthoptera, they are not well represented in the tropics. Most of the more than one thousand species live in high latitudes.

Stonefly-like insects are well represented in the Kansas Permian by an ancestral extinct order, the Protoperlaria. These ancient "stoneflies" had, like some other extinct orders, the short, wing-like lobes on the pronotum. Their young were obviously aquatic, with swimming legs and several pairs of lateral tracheal gills on the abdomen. Of the living stoneflies, the Australian family Eustheniidae is most like the extinct Protoperlaria.

Stonefly nymphs live in flowing, well-aerated water, where they shelter under stones or in debris. Fringes of long hairs on the legs provide swimming surfaces.

The adult stonefly is soft, weak-flying, and tender jawed. Apparently some do not eat; others nibble on algae crusting rocks and trees or on other vegetation. The geographic distribution of stoneflies shows that they like cool climates. Many species carry this partiality for cold to an extreme: the nymph grows little during the summer, then quickly matures in cold weather, and the adult emerges to fly and mate in midwinter.

Two of the most significant structural features of the adult are characteristic also of the Orthoptera—the hind wings, with a fan-like area, and the cerci (Fig. 49). Some important differences include, of course, the divergence in way of life between young and adult and the structure of the front wings. In the relatively sedentary and short-lived stoneflies, the front wings do not form thickened, protective wing covers. The stoneflies are able to run, however, and, correlated with this, they fold the wings flat over the abdomen. The wings are not coupled in flight. Females of many widely different groups of insects are wingless, the males winged, but some of the stoneflies are unusual in that they turn the relationship around; it is the males that are flightless.

When the young of the mayflies (order Ephemerida) became specialized for life in the water, they took over nearly the whole life cycle: the fragile mature insects, the ephemera—*Eintagsfliegen*, the Germans call them—live only hours or days.

The past of the mayflies is firmly rooted in the "first families" of winged insects. They are related to three extinct orders that lived in Carboniferous times: the Palaeodictyoptera, Protephemerida, and Megasecoptera. It may be that some of these were aquatic, although the only immature specimens that have been found (Megasecoptera) were not obviously specialized for aquatic life. The Palaeodictyoptera are the insects usually pictured in discussions on the origin of wings, because they had the pronotal lobes—possible remnants of ancient gliding planes—well developed. Because the few complete specimens that have been found had the wings spread out flat, it is assumed that they were, like the mayflies, unable to flex the wings. Some of the Megasecoptera probably could flex the wings. Both the Palaeodictyoptera and Megasecoptera were important groups in Carboniferous and Permian times, but the order Protephemerida is known from only a single species, of Upper Carboniferous age. It seems to be intermediate between the Palaeodictyoptera and the true mayflies. True mayflies first appear in Permian rocks and are abundant in the Kansas locality.

The Permian mayflies differ from those living in that the two pairs of wings are about equal in size and had no coupling mechanism.

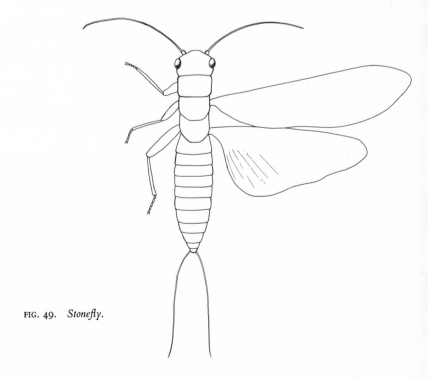

FIG. 49. *Stonefly.*

Modern mayflies have small hind wings (Fig. 50)—they have been eliminated in some—and the two pairs are coupled in flight. Some of the mayflies have a peculiar dancing flight in which they rise and fall slowly. These are unable to make much headway in forward flight. Their wings are corrugated, which would give the wings great rigidity if it were not for the fact that some of the veins lying on the ridges have each a large weak spot that causes the wing to collapse partially on the upstroke. This arrangement seems to be adapted to their up and down flight. Those mayflies skilled in direct forward flight have wings of the ordinary type, nearly flat, and with the veins not weakened.

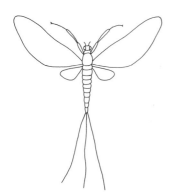

FIG. 50.   *Mayfly.*

The mayflies possess some remarkably primitive features. One of these is the triple set of tails. Only the wingless thysanurans, presumably near the ancestry of winged insects, also have them. The lateral tails are elongate cerci; the central one is said to be a drawn-out portion of the tergum of the last body segment. In the adult mayfly the long, streaming tails probably have something to do with the aerodynamics of flight, in addition to any sensory function they might have.

Another supposedly primitive characteristic of mayflies is the fact that they molt once after the wings reach full size. It is thought that extinct orders of primitive winged insects may have continued to molt indefinitely after becoming mature, in the manner of the Thysanura, but there is no direct evidence that this is so. The extra winged stage in the mayfly life cycle is called the subimago. It may be that this extra stage is a specialization to get the insect out of the water safely. The skin of the subimago is closely set with fine hairs, which may waterproof it enough that the insect can fly up from the surface without getting bogged down. After flying some distance away, the subimago settles down and molts, leaving a cast of a perfect winged insect behind. These are the only cast skins with full-sized wings that one will see, since no other winged insect molts after becoming winged.

Finally, the jaws of mayfly nymphs are remarkably primitive. Like those of some thysanurans, they have only a single point of articulation with the head. The jaws of other insects, when of a sturdy chewing type, are strongly articulated on two points.

Mayfly nymphs are probably the best adapted to life in the water of all major groups of aquatic insects, in the sense that they are varied, with different groups of mayflies being intensively adapted for one or another kind of aquatic environment. Some, like the well-known *Hexagenia*, burrow in mud, and their whole anatomy is modified to correspond to this mode of life. Others are flattened discs and cling sucker-like to stones in swift currents. A species of East Africa burrows in wood, lines its tube with silk, and has a complex filtering device for sieving out microorganisms. Apparently nearly all are herbivores. The tracheal gills are paired and are on the abdomen; they often beat rhythmically to produce a respiratory current.

Nymphal life lasts one or two years. The nymph at maturity comes to the surface, and the subimago climbs out of the nymphal skin. Mayflies of a given species emerge synchronously and sometimes appear in incredible numbers, gathering at city lights to pile in drifts several feet deep. On large trout streams of Michigan the two-inch mayfly *Hexagenia* moves upstream in packs that extend from bank to bank, flying with a loud rustling noise. Trout feed avidly during these hatches; many trout lures imitate the mayflies. During the mating flight, the male typically approaches the female from below. Correspondingly, the upper half of the compound eye usually differs from the rest, and this modified part is sometimes set out on a pedestal (Fig. 51).

The familiar dragonfly (order Odonata) is the most powerful insect of the skies; it lives in the air and feeds only on insects caught on the wing. Surprisingly, the dragonfly is not a recent product of insect evolution, but is one of the most ancient types known. In its long and isolated history, it became and has remained the most distinctive of all orders.

The huge ancestral dragonflies (order Protodonata) that lived in the late Paleozoic, with wingspans of twenty-six to thirty inches, must have been in their time the rulers of the air and probably could fly forty or fifty miles an hour, about twice as fast as any living insect. It was to be tens of millions of years before the birds came on the scene. These large insects must have had poor maneuverability compared to birds, and, although it is not known when birds with the speed and dash of modern falcons and hawks came into existence, or

FIG. 51.   *The compound eye of this remarkable mayfly from Peru is divided into two parts: the part in the usual position on the side of the head, and a coarsely faceted portion on a pedestal on top of the head. This male presumably uses the extra compound eye in locating the female during the vertical dance of the mating flight. (After Edmunds.)*

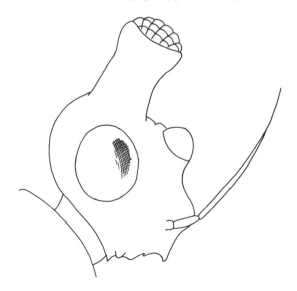

even when the giant dragonflies disappeared, it is possible that birds drove these insects into extinction. The larger of the modern dragonflies span four or five inches; such modest-sized dragonflies also were contemporaries of the giant ones.

The flight mechanism of dragonflies is unique. In order that the turbulence produced by the front wings may not reduce the efficiency of the hind pair, the front and hind pair beat alternately. Also, the thorax may be so altered that the hind pair operate in an area below, rather than behind the front wings. The muscles move the wings on a principle different from that used in all other insects. The large flight muscles extend from notum to sternum. The central group by its contraction depresses the notum, which elevates the wings. This is the same mechanism as that in other insects, but the lateral group is attached directly to the wing base and force the wing down. Other insects do this with longitudinal muscles that arch the notum.

The dragonfly scoops its prey out of the air with the bristly front legs, which it holds under the thorax to form a basket. With its sharp jaws and maxillae it chops up the prey, hide and all, into a fine hash.

It is impressive to see a large dragonfly loitering in air nearby, with its wings beating as steadily as if driven by a motor, only now and then showing a burst of greater amplitude as if to demonstrate that it is holding power in reserve. The slender abdomen projects stiffly behind the massive thorax that houses the muscles of flight. Huge hemispherical eyes, filled with clouded reflections, cover most of the head; should they detect a movement, the idling flight mechanism is swiftly thrown into gear, and with a rustle of wings the insect swerves away after its victim. They are fond of deer flies, which swarm in open meadows near woodlots, and such meadows are favorite hunting grounds for the dragonflies, even though far from water.

In a spacious mountain canyon one may see the sun glinting on the wings of hundreds of dragonflies, from half a mile away, as two or three hundred feet high in the air, each patrols its bit of airspace. They are in size well below the size range of swifts and swallows, with whom they compete for the supply of drifting and flying insects, and probably are themselves relatively immune from attack by these

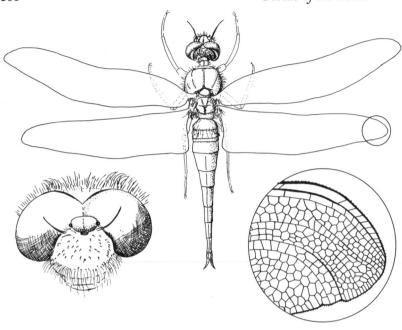

FIG. 52.   *Dragonfly. Front view of head (lower left).*

birds. Although swift and precise in their flight, the birds lack the short-range maneuverability of the smaller dragonflies, so that these insects and the swifts and swallows range the skies together, the most exquisitely adapted of their groups for an aerial existence.

Adaptation to life in the air extends to mating. The copulating pair flies as swiftly as the single insect. In these "tandems," both male and female fly forward, and in such relative positions that all four pairs of wings operate without interference. The male pulls the female along by the head or thorax, which he grasps by forceps at the end of his abdomen, and the female hangs on by curving her abdomen under and forward to fasten the tip to the genitalia of the male, which are on the underside of the abdomen far forward, near the thorax. Before mating, the male has charged the genitalia with sperm from the reproductive opening, which is situated, as in other insects, at the tip of the abdomen.

The aquatic nymph of the dragonfly also is a predator. It is easy to get one to feed (a few Daphnia and a nymph in a bowl of water quickly gets results), although the capture is so nearly instantaneous that the method is at first hard to make out. The nymph shoots out at lightning speed its labium, a long, extensible arm carried folded under the head and furnished with a pair of grasping hooks (Fig. 53). When pulled back, the labium serves as a kind of bowl in which the meal is devoured at leisure.

Dragonfly nymphs live in varied aquatic habitats, including warm and stagnant waters, and the group is very well represented in the tropics, where most of the four or five thousand species live. Like the stoneflies and mayflies, they breathe with tracheal gills. In one group the gills are in the rectum, and, by contracting and expanding

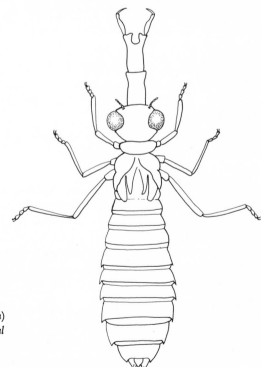

FIG. 53. *Immature stage (nymph) of dragonfly, with the raptorial labium extended.*

this cavity the insect creates the respiratory current that aerates the gills. This serves also for fast swimming, since the insect jets ahead by spurting water forcibly out of the respiratory cavity. The other group of dragonfly nymphs have three conspicuous tails that are the tracheal gills; these insects also absorb oxygen through the skin and the walls of the rectum.

These two groups of dragonflies are, respectively, the dragonflies in the restricted sense and the damselflies. True dragonflies are larger and more robust and fly more powerfully. Their hind wings are broadened at the base, hence the name Anisoptera—the two pairs of wings are not alike. The wings of the damselflies or Zygoptera are very narrow at the base and are held vertically over the body, as in mayflies. Despite their delicate build and weak flight, the damselflies also are predators.

On the island of Oahu the nymph of a species of damselfly has successfully invaded the terrestrial environment. It lives in the damp litter under ferns. There has been little morphological change in its evolution from related damselflies, which live in small basins of water held by plant leaves. Hairs on the tracheal gills apparently hold enough moisture to keep them damp and functioning. Perhaps this is the kind of step, small in itself, but great in its potentialities, that has founded important new groups of insects in the past. On the Hawaiian Islands there were—before the advent of man and the animals, large and small, that he brought with him—few predators and competitors to interfere with poorly prepared experiments of this kind. Given a few million years, this damselfly might produce, or might have, except for human interference, a new order of terrestrial insects related to the dragonflies.

The order Neuroptera stands in sharp contrast with the preceding three orders (stoneflies, mayflies, and dragonflies) because it has complete metamorphosis. Quite possibly the Neuroptera were the first, and perhaps even the only group of insects to invent the true larval stage, where there are no external traces of wings. The aquatic alder flies are judged to be the most primitive of the Neuroptera, which would indicate that the larval stage first appeared in insects whose young lived in the water.

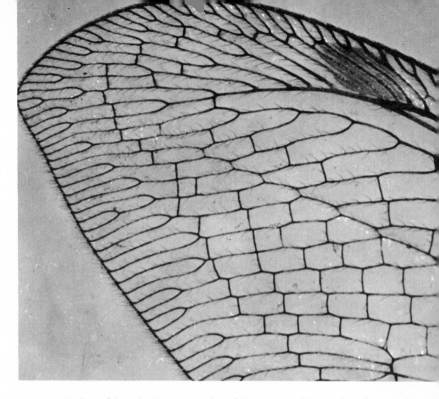

FIG. 54. *Members of the order Neuroptera, that of the net-winged insects, have large, rather fragile wings with intricate patterns of venation as shown above.*

It is believed that the Neuroptera gave rise to all the groups with complete metamorphosis, and these comprise all the orders yet to be discussed—Neuroptera, Coleoptera, Lepidoptera, Diptera, and Hymenoptera as well as some minor orders. They include the vast majority of the species of insects now living. The orders with complete metamorphosis are sometimes termed the neuropteroid complex.

The three divisions of the order Neuroptera—the alder flies (Suborder Megaloptera), the flat-winged neuropterans (Planipennia), and the snake flies (Raphidiodea), totaling about three thousand living species—are remnants of groups once abundant and diverse, as shown by the record in Permian rocks. Sometimes these three groups are ranked as orders.

Alder flies are the most primitive of the three. They look much like stoneflies, although they lack the cerci of that order. The wings have a small area along the hind margin that folds like a fan, in a way reminiscent of the stoneflies and Orthoptera. The larvae live in the water, are predators, and are equipped with biting jaws. Like the primitive aquatics, they breathe with tracheal gills. In some the gills resemble jointed legs, a pair to each abdominal segment; in others these are supplemented with tufts of fine filaments at the bases of the leg-like structures. It is interesting that these aquatic larvae emerge from the water to pupate, seeking shelter in soil, moss, or under logs.

The best known North American species of the suborder Megaloptera is *Corydalus cornutus*, and this is known chiefly for its larva, the hellgramite. Fishermen regard this tough-skinned, three-inch, ferocious-looking larva as excellent bait for trout and bass and hunt them by turning over stones in the beds of swift streams, allowing the insects to drift into a net held below. The adult, called a dobson fly, is a remarkable insect, not often seen but easily remembered, with a wingspan of five inches and, in the male, with huge forward-projecting jaws that may be half the length of the body.

The snake flies, with fewer than a hundred species living, are strange-looking insects, with the prothorax lengthened out into a long neck (in German, *Kamelhalsfliegen*). Also, the female is unique among the neuropteroids in having a long, sword-shaped ovipositor. The larvae are terrestrial and predaceous and, like the Megalopterans, have ordinary biting mandibles. All the North American species occur in the western half of the continent.

Planipennia, or the Neuroptera of many recent classifications, are today, with their two thousand species, the dominant neuropterans. The predaceous larvae do not eat solid food but drink the body fluids of their prey through the grooved mandibles. The groove on each long, sharp jaw is closed over by a slender part of the maxilla. All are terrestrial, except larvae of the peculiar sponge fly, *Sisyra*, which eat fresh water sponges. The Planipennia are a varied group in warm regions, with many large and sometimes bizarre types. The two dominant families are well represented in North America—the Chrysopidae (golden-eyes, lacewings, or aphis lions) and Myrmeleonidae (ant lions).

Golden-eyes, like the ladybird beetles, are the gardener's friend.

The larva, or aphis lion, feeds on aphids, lifting one up in its long jaws, draining it, flipping the empty hulk back with a toss of the head, and straightway advancing to another. Adult lacewings are gauzy-winged brown or green insects nearly an inch long (Fig. 55). In life their eyes often are bright gold. Some species emit a remarkably powerful nauseating odor.

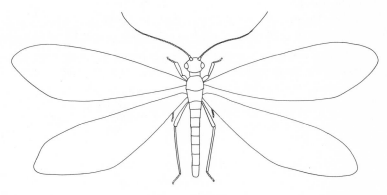

FIG. 55. *Lacewing.*

Most of the species of the now dominant family of Neuroptera (the ant lions, or Myrmeleonidae) make their living by digging pitfalls designed to trap small insects. These conical pits are dug in dust or sand, usually in a location sheltered from rain. The squat larva dug up from the bottom of the pit may be placed in a can filled with dry sand, where it digs a new pitfall and accepts prey given it. When the ant or other insect tumbles in, the ant lion pelts it with puffs of sand as it tries to scramble out of the crater. Adult myrmeleonids, weak flyers that come out mostly at dusk, resemble damselflies in form and size.

A minor group closely related to the Neuroptera is the order Mecoptera, that of the scorpion flies. In late Paleozoic times these were among the three most important orders of insects, judging from the fossil record. They are now a remnant of some three hundred species, but in damp temperate forests they may be locally the most common large insect to be found.

Except for some Australian species, the scorpion flies have the head prolonged vertically into a beak, which bears the chewing mouthparts at the tip. The larvae, with possibly an exception in Australia, are terrestrial.

The common scorpion flies fall into two groups. One, typified by *Panorpa*, are active, rather wasp-like insects that run over foliage looking for insect prey. The male has a bulbous enlargement at the tip of the abdomen that gives the group its name.

Another group, represented by *Bittacus*, is one of sluggish, crane-fly-like insects. These spend much of their time hanging from leaves by the front legs and capture with the long, dangling hind legs insects that might blunder into them or else pounce upon the prey. Although these insects are relics of an earlier age, they fairly swarm in some local areas of the hardwood forests of the eastern United States.

17

Beetles

Most beetles live in concealed places—under stones, logs, loose bark, but so attractive are their complexly sculptured forms and their gem-like colors that they have long been sought after by collectors and are relatively the best known of the great orders. About one out of every four described species of insect is a beetle—two hundred and fifty thousand species in all.

One of the fraternity of beetle collectors was Charles Darwin who, while a rather unwilling college student, collected beetles for sport, admiring their beauty and trying for records in rarity. He recalls that in one especially good collecting place, no sooner had he picked up one rare beetle than he saw another and took it with his free hand, only to be confronted with yet a third prize specimen. After briefly weighing the possibilities he threw one of the beetles in his mouth and remembers that it was one of the kinds fitted with chemical defenses, giving it a fiery taste.

Beetles, which comprise the order Coleoptera, owe their success largely to the structural features of the body which fit them for living in the numerous and safe cryptic habitats in the world. The fragile hind wings usually are large and effective instruments of flight, but when not in use, they can be folded under the hardened, cup-shaped front wings, called elytra, or wing covers, where they are protected from abrasion, and allow the beetle to move freely about (Fig. 56).

So compact is the body form and so short the wing covers that usually the flight wings have to be folded not only in lengthwise folds but also in a transverse fold for storage.

With the beetles we take up the first of the "Big Four" of the insect orders—Coleoptera, Lepidoptera, Diptera, and Hymenoptera—which contain between them the overwhelming majority of the species of insects now living. The members of these four orders are exceedingly diverse in structure, but they have one thing in common: complete metamorphosis, in which the young differ much from the adult. Presumably it is this division of labor within the life cycle of the insect—the immature, or larval, stage usually devoted to feeding, the adult stage to reproduction, dispersal, finding suitable habitats for the young, or even caring for the young—that has accounted in large part for the outstanding success of these orders.

As we saw in the preceding chapter, the Neuroptera may have been the first of the insect orders to achieve complete metamorphosis. Whether correctly or not, it is usually assumed that this invention was made only once and that all the orders with a larval stage are

FIG. 56. *Buprestid beetle. The wing cover and flight wing are shown extended on the left side, while part of the folded right flight wing protrudes from under its wing cover.*

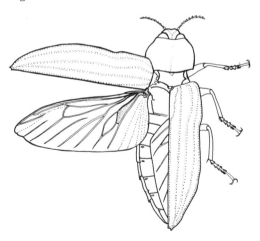

descendants of some Neuroptera-like group. If this concept of the origin of the larval stage is correct, it may be that the specialization of the young for life in the water was the starting point for the extreme differentiation between young and adult that occurs in complete metamorphosis.

Beetles usually are not very good fliers; at least they do not have the precise, darting movements of some of the flies, bees, or wasps. Their flight usually is heavily direct, as one sees at lights in the evening, when the straight-line flight of the large June beetles terminates in an unchecked thud against the screen. A dispersal flight, which assures that the species occupies all available habitats in the area, is characteristic of many species of beetles and usually is made at night when the air is full of these unseen wanderings. Sometimes the magnitude of these dispersal flights becomes known after unusual weather conditions, such as when a very strong offshore wind at night blows the insects far out onto a large lake, and a succeeding inshore wind drifts them in again. On the beaches of the Great Lakes are sometimes windrows of drowned beetles miles in extent, with millions of individuals and dozens or even hundreds of species. Some beetles carry out this straight-line, distance-consuming dispersal flight in daylight, and a characteristic sight on early spring days, when the sun is low in the sky, is the glint on the wings of the thousands of beetles that are surprisingly closely spaced in the air as far as one can see.

The wing covers are held outstretched in flight, either stationary or vibrating; just what their contribution is to flight is not known, but the minority of beetles that have unusually good powers of flight do away with this apparently clumsy arrangement in one way or another.

The thousands of species of staphylinid beetles have reduced the wing covers to rather small scales which offer little impediment in flight. These beetles fly in a characteristic upright pose, something like that of a sea horse, and some can hover and shift direction quickly. The hind wing is so large relative to the miniature wing cover that they have to be folded in a complicated manner into a thick pad when put in storage.

A group of the scarab beetles, called the cetonines, also are expert fliers and in flight are almost indistinguishable from some of the swift bees. They fly in the daytime and sometimes visit flowers. As they hover, it is possible to observe that the wing covers are held smoothly back over the body, instead of outstretched. A close look at a specimen in the hand shows that this is made possible by a relatively simple device, a notch on the margin of the wing cover which allows the flight wings to protrude and move without hindrance.

The more primitive—judging from their structure—of the beetles are carnivores, both as larvae and adults. Most active and dashing of these are the tiger beetles, brightly colored, long-legged creatures that run swiftly over open ground in search of their insect prey. They take wing when pursued by the collector, flying fast for a short distance, the change of pace being so abrupt that they are easily lost to sight. The large, bulging eyes and long, scimitar-shaped jaws help fit them for their mode of life.

Although also predators, the young of the tiger beetles are completely sedentary, lying in wait for their prey. They live in vertical burrows, with the entrance neatly plugged by the head and front part of the thorax. When an unsuspecting insect walks over the plug, it is seized by the sharp mandibles. Bracing itself in the burrow with suitably contrived hooks on its back, the beetle larva hauls the struggling prey down into the burrow to devour it. When an observer first intrudes onto a hunting ground where tiger beetles are common, he sees that the ground is peppered with holes the size of a matchstick and smaller. If he sits and waits for a time, these holes disappear one by one, as the beetle larvae come back to resume their stations at the burrow entrances.

The most abundant, both in individuals and species, of the primitive carnivorous beetles are the carabids, or ground beetles (Fig. 57). Most of the several thousand species are small plain black or brown beetles that live on the ground, spending much of their time hidden under fallen wood, forest litter, or under stones.

The collector will notice that many of them emit strong odors, ranging from fragrant to unpleasant, when picked up, odors reminiscent of the esters of alcohols encountered in the biochemistry laboratory. Yet other carabids eject corrosive liquids that burn the skin.

Such a beetle is the green and red bombardier beetle (*Brachinus*), which emits from the tip of the abdomen a volatile liquid that vaporizes with a miniature explosion into a smoky cloud.

Some of the larger ground beetles feed on snails. These have part of the thorax lengthened out into a neck that enables the beetle to work its way well back into the snail shell. Another large carabid, *Calosoma sycophanta*, about an inch long, has been imported from Europe into North America to combat the gypsy moth. One of these voracious beetles may devour as many as five hundred moth caterpillars during its life. Adults of some of these large carabid beetles may live for several years, a life span quite unusual among adult insects.

Closely related to the land-dwelling carabid beetles are the diving beetles, of the family Dytiscidae, all of which live in fresh water. Diving beetles obviously are remodeled land beetles, with a tracheal system for breathing air—the insect has to come to the surface periodically to get it—and the larva crawls out of the water to transform into the pupa and the adult stage.

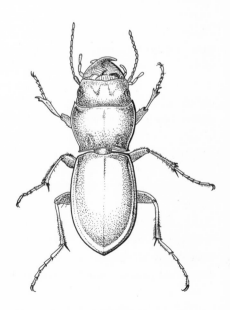

FIG. 57. *Carabid, or ground beetle.*

Diving beetles are among the largest and most active of insect predators in the water. Reaching a length of over an inch, they are able to prey on tadpoles and small fish as well as on other insects. The larva, called a water tiger, overcomes prey with the aid of its very long, needle-sharp jaws. These are hollow, and through them powerful digestive enzymes are pumped into the victim, which kill it, then liquefy the body contents, which the water tiger drinks through the jaws, as through a pair of drinking straws.

The adult beetle is equipped with powerful hind swimming legs, which make it one of the most expert swimmers among aquatic insects. As in many aquatic insects, the dispersal flight is an important event in the life of the beetle, for by this means widely scattered ponds and streams are populated by the most species. Most diving beetles have a dispersal flight and fly fast and far at night. So alert are they for any evidence of water that when they see the moonlight gleaming on the polished top of an automobile or the glass roof of a greenhouse, they sometimes slant down to crash into the hard surface.

Also related to the carabid or ground beetles are the gyrinid beetles, gleaming bronze or black insects that swim, closely pressed to the surface of the water, at incredible speeds, pushing up a long V-shaped wake. Except when frightened, they usually swim in crowds in a maze of circles, hence the popular name whirligig beetle. The remarkably wide, flat legs give them their swimming ability. In an aquarium they continue to swim at breakneck speed, skillfully avoiding the sides. One observer dabbed them with luminous paint and watched them swim in the dark, still avoiding obstacles. When he removed their antennae, the ability was lost, leading to the conclusion that the antennae detect pressure waves reflected off obstacles. The compound eyes each are divided, so that the beetle really has four eyes, one pair looking upward, one down into the water. Sometimes they gather in great swarms on the water surface, and, when large numbers are scooped up in the net, the strong, not unpleasant odor of a defensive secretion is noticeable. Although the beetle spends much time on the surface, it is able to dive. The larva is completely aquatic, being furnished with tracheal gills, and it spins a cocoon for the pupa underwater.

The family name of the Hydrophilidae (from water-lover) is not entirely appropriate, since many of these beetles never go near the water; but nevertheless some of the most familiar and abundant of the water beetles do belong to this family. It is a somewhat isolated group, not related to the ground beetles. The adults usually eat plant materials, and they are poor swimmers compared to those active predators, the diving beetles. On the underside of the beetle is usually a single, long, ski-shaped structure, which looks as if it would help in sliding over the bottom while feeding on decaying vegetable ooze. What look and function like antennae are actually the palpi of the maxillae, the antennae being used in part to form a funnel that leads air from the surface down to a flat, glistening bubble on the underside of the abdomen.

A person reading beside an open window on a warm summer evening will be visited by a select group of minute insects that can get through the window screen. Among these are some tiny beetles that alight on the book, twist their sharp-pointed abdomen about in movements that seem to help fold the wings, run about expectantly for awhile before they discover that this is not the place, and take off again. These are members of the family Staphylinidae, an immense and little-known group of about twenty thousand species. Although the biologies of most species are unknown, probably the group is primarily carnivorous. Some of the large and best known species live in carcasses, but even these probably feed on other insects that they find in the decaying mass.

Hundreds of species of staphylinid beetles, themselves more or less ant-like in appearance, live in ant colonies, where they are tolerated —even prized—because they secrete a substance which is eagerly sought after and licked up by the ants. Probably the substance resembles the "social hormones" secreted by the ants themselves, which seem to make the ants prize each other's company and even to relish the never-ending task of feeding their young. The beetles have seized upon this foothold and expanded it to the degree that they live lives of leisure in the colony, with the beetle now and then caressing an ant in such a way that the ant disgorges food for it, with the ants taking care of the young beetles, and with the beetle supplementing its diet by eating young ants, no doubt with its hosts looking on indulgently. These beetles obviously are genuine ant-lovers; hence they are termed myrmicophiles.

The secretions of other, free-living staphylinids are often malodorous or irritating and serve to protect them from their enemies. These secretions give to the tiny beetles that sometimes get in our eyes a fiery sting.

The secretion of yet another group of staphylinids (*Stenodus*) is put to use in swimming. Exuded at the tail end of the animal, it lowers the surface tension of the water, allowing the stronger tension of the water film in front to draw the insect forward. This is the principle of the toy camphor boat. A biologist who lived as a youth in China recalled that he used to find these beetles in numbers along the banks of streams. When thrown into the water, they immediately made for shore on the surface with a steady and swift movement which seemed to him mysterious—he could never make out how they did it.

Burying beetles and carrion beetles (family Silphidae) can usually be obtained in a few days by exposing a small dead animal, which should be tied down or screened to keep large animals from taking it. These beetles fly in, apparently from considerable distances, guided by scent. The two most common types are large beetles, over half an inch long: *Silpha*, flat in form and dull colored, and the bright orange and black *Nicrophorus*, with rather short wing covers. Species of the latter genus are the burying beetles. When the adult beetles find the carcass, they are said to dig dirt out from underneath it, so that it gradually sinks into the ground and is eventually completely covered. This would keep the food supply from drying out and perhaps protect it from larger scavengers.

Ladybird beetles (Coccinellidae) appear to have few endowments other than a good appetite to fit them for life as predators, but since they eat mostly such soft and slow animals as plant lice, scale insects, and mites, they require little in the way of speed, armor, or weapons. Apparently they are well defended from their enemies by the bitter and even poisonous qualities of their yellowish blood, which exudes at the leg joints when they are picked up. The most familiar of these hemisphere-shaped beetles are brightly colored, which advertises their distasteful properties.

Observations on one species of ladybird showed that the larva ate from two to five hundred aphids to reach maturity and that the adult beetle continued to eat them. Their voracity, coupled with their short life cycle—sometimes no longer than a month—makes them important in the control of insect pests. Various species are grown in entomological laboratories and sold to agriculturalists for release in their fields. The habit of some species of hibernating in great masses, often on mountaintops, makes it possible to harvest enough of them to sell.

A few of the ladybirds are plant-eaters, and one of them, the bean beetle (*Epilachna*), is an important pest.

The Meloidae, or blister beetles, are plant-eaters when adult, but their larvae are primarily carnivores, although some larvae have shifted to a diet of pollen, which nutritionally is much like animal protein. The young live at the expense of two other groups of insects, grasshoppers and bees, which are quite unrelated but have the habit in common of nesting in the ground. The newly hatched meloid larvae, called triungulins, are minute, long-legged, and incessantly active. Those that are enemies of grasshoppers run about until they find a cache of grasshopper eggs, which they devour. With each molt the young beetle changes greatly in appearance, each time becoming more grub-like and helpless. The adult beetle emerges from the earth to take up its vegetarian existence.

The triungulins of the meloids that victimize bees do not search out the nest but climb up on the tips of plants, often coming to rest on flowers, where they wait for flower-visiting insects. When one alights, they swarm onto it and cling to the hairs. The triungulin fortunate enough to have chosen not only a bee but a bee of the appropriate species then rides to the nest where it will be able to complete its development. After finding the store of pollen and nectar in the nest, the triungulin kills and eats the egg or young larva of the bee, then devours the stored food. As it grows, the larva becomes more helpless, and, should another triungulin come along, the older larva shares the fate of the original owner of the pollen mass.

The biological significance of variations in nesting behavior of the bees is shown by the effect of these differences on the meloid parasite. In one example provided by a pair of species of bees, the vulnerable species lays its egg on the side of the nest, where the triungulin can

easily find it, whereas the immune species lays its egg on the pollen
mass and surrounds the egg with a moat of honey that the triungulin
can not cross.

Adult meloids differ from most beetles in having soft elytra. They
may be conspicuously colored, are slow moving, and feed in exposed
places. The colors of the meloids are sometimes flashed when the
beetle is disturbed; one black California species spreads its wings and
walks in a peculiar jerky fashion, displaying red stripes on the abdomen.
Such warning displays advertise that the beetles have good chemical
defenses. The chemical involved is the blistering irritant cantharidin,
which is concentrated in the wing covers of some species but also
occurs throughout the body and may be emitted in droplets from the
leg joints. The exudation of some of these beetles is immediately
painful to human beings. The Spanish fly (*Lytta vesicatoria*) is a
European meloid beetle, brilliantly blue and green, that furnishes the

FIG. 58. *Fragility and armored strength in the insect world contrast in this pair of South
American insects: a giant scarab beetle, which measures four inches, and a transparent-winged
satyrid butterfly.*

cantharidin of pharmacology. This quite poisonous substance—the lethal dose for man is about 0.03 grams—has been used as an aphrodisiac.

The incredible fireflies are, like the meloids, soft-skinned beetles with carnivorous larvae. The cold light of the firefly comes from the underside of the abdomen, where there is an area of transparent skin, a group of living light-producing cells, and a layer of reflecting material. The glowing cells are filled with particles once thought to be luminescent bacteria but now considered to be mitochondria, which are the energy-generating centers of all living cells. In brief, the energy extracted from nutrients by the mitochondria is ultimately released as radiant energy of short wave length, rather than as heat. The firefly is able to turn its light off and on at will, and the frequency of the flashes and the color of the light is characteristic of the species. A male is attracted to a flashing female in a glass vial, but not to a nonflashing one. Males also will not come to a female hidden in a porous box, so it is light, not odor, that is the attractant. It is said to be possible to call male fireflies to a flashlight that is properly operated. One entomologist writes that the fireflies glow also for pure enjoyment. In some the female is wingless and has a very bright light. In our familiar fireflies, both males and females have lights.

There are about thirty thousand species of scarab beetles (family Scarabaeidae), ranging from minute polished hemispheres barely a millimeter long to the six-inch Hercules beetle and from dingy brown beetles to insects that look as if they had been cast in twenty-four-carat gold (see Fig. 58).

Scarab beetles are diggers and are clumsy afoot. Their front legs are wide, spurred, and heavily muscled. Scarabs have a larger sensory surface on the antennae than is usual among beetles: the last few segments are expanded laterally into thin plates, or lamellae, which are usually held together in a compact club and are spread widely when the antennae are in use. The larvae, which are scavengers or vegetarians, live in earth or rotten wood. They are pale, soft-bodied, and bent in a semicircle about a flabby paunch. Large numbers of these white grubs live in turf, where they eat grass roots.

A well-known plant-eating scarab is the Japanese beetle (*Popillia japonica*), which was brought into the eastern United States in the early 1900s. The beetle lays its eggs underground, in turf. At first the young eats soft, decaying plant material, but soon it shifts to live roots and may kill the grass. The large grub overwinters in the ground, feeds some more in the spring, then transforms into the adult beetle, which emerges in June or July. Most crop damage is inflicted by the adult, which feeds on over two hundred species of plants, attacking both fruit and foliage. So serious a pest is it that a good deal of effort has been directed toward controlling it. One effective method of killing the grubs is to infect them with a bacterium that causes "milky disease." Laboratories prepare dried spores of this bacterium in quantity by injecting spores into grubs, allowing the disease to take its course, then grinding up the sick grubs with mineral powders. An ounce of the resulting powder, which is sold commercially, contains nearly three billion spores, somewhat more than the yield from a single dying grub.

The common May or June beetles (*Phyllophaga*), also vegetarians, are the brown beetles that swarm about lights on warm evenings of early spring.

The varied dung beetles are scarabs that dig underground nests and stock them with dung for themselves or their grubs. Some species dig the nest under the dung; others, the tumblebugs, roll the supply, sometimes a ball the size of a fist, from a distance, pushing it along with the hind legs. The scarab may lay only a very few eggs in this mass of food and tends the young to maturity. An Indian scarab is said to cover its dung ball with clay; when some of these were dug up, they were thought to be ancient stone cannon balls.

Three families of vegetarians together make up a large fraction of all beetles: the leaf beetles (family Chrysomelidae); longicorn beetles (Cerambycidae); and weevils (Curculionidae), which total nearly one hundred thousand species. The first two families are closely related and grade into each other; the weevils may be related to the other two. They have in common a tarsal structure (in the adult stage) which seems to be adapted for clinging to foliage—wide, flattened tarsal segments, thickly padded underneath with velvety hairs.

The typical leaf beetle is a squat, nearly hemispherical insect with bright colors, but there are many departures from this type in this huge family of twenty-five thousand species. These beetles feed openly on leaves. To escape enemies, they fold up their legs and fall to the ground at the slightest disturbance to the plant; some do not require even the mechanical stimulus but drop when they see a movement. Others cling phlegmatically to their leaf, and these presumably have the merit of being poorly flavored. Perhaps the most familiar chrysomelids are the Colorado potato beetle (*Leptinotarsa decemlineata*) and the asparagus beetle (*Criocerus asparagi*), the latter an attractively colored blue, red, and yellow beetle that leaves its eggs abundantly on asparagus shoots. A strikingly beautiful, gleaming blue species that often attracts attention is *Chrysochus auratus*, which is common on dogbane. The common tortoise beetles, which cling tightly to their leaf like an inverted saucer, relying on armor protection, are brilliantly gold in life, but the color disappears at death. Larvae of the leaf beetles are, like the adults, foliage-eaters.

Longicorn beetles, named for their long antennae, as larvae specialize in eating wood, although some mine in green stems. The adults often feed on pollen and nectar. Among the longicorns are adept mimics, some with short wing covers imitating Hemiptera; others, brightly patterned and with quick nervous movements, resemble Hymenoptera. Yet others are concealingly colored. The stolid red and black milkweed beetle (*Tetraopes*) presumably tastes bad as a result of its confirmed habit of eating milkweeds. The longicorns, which are of larger than average size and which have great diversity of color and pattern among the twenty thousand species, are favorites with beetle collectors.

The beetles as a group are without ovipositors, but the weevils made up for this deficiency when they evolved their snouts, which are used primarily for boring holes in plant tissues, holes in which the eggs are placed. Some have snouts as long as the rest of the body and can gnaw deeply into buds, seeds, and fruits. The snout is a prolongation of the head that carries the antennae part way with it and bears at the tip the chewing mouthparts. Sometimes the snout is grooved to get the long basal piece of the antennae out of the way when the snout is in use. The larva that grows in the midst of the food where the egg was placed by its long-beaked parent is a legless, eyeless grub, more degenerate than is usual among the beetles.

About fifty thousand species have been described; and a world authority on the group states that, at the rate new species of weevils are being discovered, there will be eventually about a quarter of a million species. Some are brilliantly colored, and the large, sluggish, conspicuous kinds are often stone-hard.

Weevils drain off an appreciable fraction of the crops that are grown. The cotton boll weevil (*Anthonomus grandis*) each year destroys in the United States from twenty to forty percent of the cotton crop. It entered Texas from farther south sometime late in the nineteenth century and within thirty years occupied most of the cotton growing areas of the Southeast. The rice weevil (*Sitophilus oryza*) and the granary weevil (*Sitophilus granarius*) are small beetles, about four millimeters long, that live in stored grain or grain products. The grub completes its development inside a kernel of grain. With four or five generations a year in the even climate of a grain elevator, these weevils can destroy an enormous amount of stored grain, and the two species are of great economic importance.

A group of beetles with habits similar to those of the longicorns, but one not at all related to them, is the family Buprestidae, the metallic wood borers. These compact, wedge-shaped beetles, a group of about eight thousand species, are sometimes brightly enameled, sometimes iridescent blue, green, or bronze, and are used in jewelry making. The larvae eat their way through wood, living or dead. Some very small species mine in the thin layer of tissue in leaves. Adults are active beetles that run quickly over timber in search of mates or egg-laying sites. Some buprestids, which lay their eggs in freshly burned timber, are attracted by smoke. They gather at the smokestacks of factories, and at a Western university they were seen coming upwind to a fire built for a monster pep rally.

The click beetles (family Elateridae) are distinguished by having a unique mechanism for jumping. If one of these beetles is placed on its back, it lies quietly for a time, feigning death, then arches its back, straightens out suddenly with a sharp click, and springs into the air. This may serve to startle its enemies, to get the beetle back on its feet, or to put distance between it and a predator. The visible parts of the jumping mechanism are a stout spine on the prothorax and a groove, for receiving the spine, on the underside of the mesothorax. Perhaps the blow of the spine against the mesothorax is transmitted to the wing covers so that they strike against the ground and hurl the insect upward. A group of elaterids that have taken up a burrowing mode of existence do not have the jumping mechanism.

Of the eight thousand species of click beetles, those of the temperate regions are usually brown or black, but there are many brightly colored ones in the tropics. Some of the tropical species are luminous, like fireflies, one in the West Indies (*Pyrophorus*) having two green lights on the thorax and a red one on the underside of the abdomen.

Larvae of many click beetles live in the soil, feeding on plant roots. These wireworms, as they are called, may severely damage crops.

With some ten thousand species, the family Tenebrionidae, the darkling beetles, is a major one, but its members, mostly black or brown beetles, are not often encountered. There are, however, two well-known ones that are common in stored grain products—*Tribolium* and *Tenebrio*. *Tribolium confusum* is a small beetle about three millimeters long that often infests kitchens, where it is found in cereal or flour. This beetle is to population studies what *Drosophila* is to genetics. It is grown in fine flour, which makes it possible to sift out eggs and larvae for easy counting, and the life cycle is conveniently short. Mealworms, or larvae of *Tenebrio molitor*, reach an inch in length. One can grow them easily in grain products, and pet shops sell them as live food for animals.

The Strepsiptera are small insects that, with the possible exception of a few rare and primitive species, are internal parasites of other insects. The best known strepsipterans (of the genus *Stylops*) infest wild bees. At maturity the female *Stylops*, a wingless, rather shapeless insect, pushes head and thorax out between the abdominal terga of her host, leaving the bulk of her body inside; this protruding, semicircular fragment is all that most entomologists have seen of a strepsipteran. By contrast, the male is a blazingly energetic winged animal (Fig. 59). During its meteoric career, which lasts hardly an hour, it flies incessantly in search of a female, which all the time is being carried about by her host. Males were rare in collections until someone learned to bait for them by tethering out large numbers of bees infested with female strepsipterans. If an unmated female—they seem to be rare—is in the lot, males gather about in a swirling cloud.

The young emerge through a genital channel that opens on the cephalothorax of the parent, and they cling to the host bee until it visits a flower, where they disembark. Here they are drunk by another

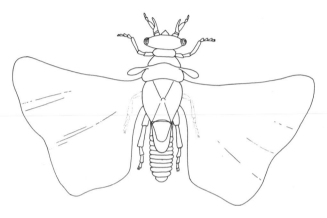

FIG. 59.   *Male strepsipteran.*

bee, or in some other way get into its mouth, and are carried to the nest. The bee regurgitates them when it stocks its nest with nectar. The strepsipterans wait until the bee larva is partly grown, then bore into it, completing their development as endoparasites. The bee is not killed by the parasite, but it often develops into an intersex that combines male and female characters.

Male strepsipterans are like beetles in that the hind wings are the flight surfaces; what correspond to the beetle wing covers are reduced to minute straps. Although sometimes considered to be a suborder of the Coleoptera, the Strepsiptera can just as well be taken as a separate order closely related to the beetles.

18

Butterflies and moths

The order Lepidoptera is for the most part one of soft-colored nocturnal insects. The bright-winged butterflies total only about ten percent of the more than one hundred thousand species.

Wings of Lepidoptera are shingled with scales, really flattened hairs, that rub off on the fingers as a fine, colored dust. This covering of hair continues over the rest of the body, and a moth may be as "furry" as a mammal (Fig. 60).

The wing area of a butterfly is large relative to body weight, giving these insects a distinctive mode of flight—slow wing beats alternated with gliding. Their flight is powerful, and some can fly hundreds of miles without rest. Moths often have wings that are comparatively narrow, approaching the usual insect type, and the wings are stroked rapidly in flight. Such large, narrow-winged moths as the sphinx moths are among the swiftest insects.

The wings are coupled in various ways. In the butterflies and some of the moths, the hind wing has a projecting "shoulder," often strengthened by a vein, that fits under the front wing. In most Lepidoptera, a long spine or cluster of bristles (the frenulum) near the base of the hind wing thrusts through a clasp on the underside of the front wing. Both frenulum and clasp are usually a group of soft bristles in the female; in the more strongly flying male the frenulum is a single stout spine, and the clasp is solid cuticle.

FIG. 60. *The head (with a compound eye below and right of center) and the back of the thorax of this large sphinx moth are densely covered with hair. This furry coat may have the function of conserving body heat for these dusk- and dark-flying insects.*

Butterflies are poor walkers and use the legs for clinging. The front pair is sometimes aborted and nonfunctional. When at rest, the butterfly holds the wings vertically, like a mayfly. Some moths are fairly agile runners, and these fold the wings back out of the way, in the manner of other running insects.

Butterflies and moths are large-eyed, as befits good fliers. The ocelli, if present, are two, usually almost hidden by "fur," and are placed near the upper margins of the compound eyes. Both the compound eyes and ocelli thus have about the same field of vision, and what it is that the compound eyes are not able to do that the ocelli do accomplish is an interesting question.

Except for a few of the most primitive moths, the adult lepidopteran lacks mandibles. The mouthparts may be vestigial, in which case the insect does not eat. When functional, they are adapted for drinking

FIG. 61. *The clubbed antennae characteristic of butterflies are shown in this series of a South American species. These butterflies are called "88s" by commercial collectors.*

nectar: two elongate pieces of the maxillae fit together to form a tube, so long that it is carried coiled in a spiral under the head.

The larvae of Lepidoptera, or caterpillars, are vegetarians, except for a minority that capture such sluggish insects as plant lice, and some—the clothes moth, for example—that feed on keratin or other animal matter. The larva of the wax-moth, *Galleria*, scavenges in bee hives and is able to digest wax. Caterpillars are equipped with powerful chewing mandibles.

Caterpillars are short-legged and slow-moving. The long, heavy abdomen is supported by up to five pairs of supernumary legs, the short prolegs, which are armed at the tip with varied patterns of small clinging hooks, the crochets. Some caterpillars have given up walking with this apparatus and progress by alternately arching and extending the body. These are the inchworms, or loopers, which include all of the extensive family Geometridae (from earth-measure) and some of the Noctuidae.

The slow-moving caterpillars defend themselves in a variety of ways. Many conceal themselves in the ground during the day and emerge only at night to feed, as do the hundreds of species of cut-worms (of the family Noctuidae). Many smaller caterpillars tie leaves into rolls that shelter them during the day. Yet others build portable cases, fashioned from bits of vegetation fastened together with silk, that they carry about in the way a hermit crab carries its shell. The tent caterpillars (*Malacosoma*) are communal and construct a dense, tangled web for their daytime roost.

The larvae of some swallowtail butterflies, when disturbed, suddenly evert a bright orange Y from the top of the thorax, which dispenses a strong unpleasant odor. Many of the furry caterpillars have the hairs filled with irritants so that they sting like nettles.

Most caterpillars pupate in a shelter or in the ground. Larvae of butterflies often pupate in exposed places, anchoring the pupa with silk lines to its resting place, where it is protected by its concealing coloration. Caterpillars of the best-known large moths spin a tough cocoon of silk in which they pupate. The moth has a large investment of time and material in the cocoon, which may weigh as much as the moth itself, this solely to protect the quiescent stage in its life cycle. Despite the invention of more or less silk-like plastics, silk is still used in large quantities. To produce twelve pounds of raw silk, growers feed thirty thousand worms one ton of mulberry leaves. Although several species of moths, in the families Saturniidae and Bombycidae, furnish usable silk, the species *Bombyx mori*, now occurring only in domestication and apparently unable to live in the wild, is the source of nearly all commercial silk.

Butterflies of some groups have evolved chemical defenses—unpleasant-tasting compounds—that give them relative immunity from predators, and these butterflies are imitated by many more-or-less edible butterflies of other groups. The mimics copy the wing pattern of the model and sometimes its mannerisms of flight.

In North America the most conspicuous example of butterfly mimicry is that of the relatively edible viceroy (*Basilarchia archippus*), which differs in color from its close relatives, and closely resembles the unrelated and quite inedible monarch. Although it has been fashionable to be skeptical about the adaptive value of this resemblance, common-sense evaluation of field experience indicated long ago that the resemblance between the two butterflies was not merely a coincidence. In recent experiments, birds given the chance to get well acquainted with monarchs would not eat viceroys, whereas birds without such experience would eat them.

A most remarkable case of mimicry is that of an African swallowtail, *Papilio dardanus*, for not one but several distasteful models. Different races of the swallowtail that live in different areas imitate the models that are common in their locality. Even in one locality the mimic has several color patterns—which may appear in the progeny of one female—each more or less resembling a protected model. Only the female is a mimic, and the male is a normal swallowtail, with tails and a black and yellow color pattern. On Madagascar is a race of *dardanus* in which both male and females are nonmimetic and resemble each other, but on the African continent the female swallowtail mimics other butterflies that are tailless and are black and white or black, white, and brown. In one locality in Africa the swallowtail mimics a species that has yellow in the pattern. In doing this it evolved a new yellow pigment that is chemically unlike the yellow of the ancestral type in Madagascar.

Although the genetic basis for each of the mimetic color patterns of this swallowtail is complex, there are single "switch" genes that elicit or repress the development of one or another pattern. In areas with a given array of model butterflies, then, selection is for certain of these master genes and against others. The species *Papilio dardanus* is apparently one of great genetic complexity and resources and one that is highly dynamic, rather than stable. The fact that this situation is expressed in the color of its wings makes it easily recognized; perhaps other widespread and abundant species of insects have a similar constitution, expressed in a more subtle fashion.

In England there are records and collections of native moths that date well back into the nineteenth century. These make it possible to study some of the evolutionary changes in the wing colors of moths. An important environmental change that has occurred during this time is connected with industrialization. Black smoke from factory chimneys not only blankets the countryside with soot but also poisons and kills lichens, which give a pale color to tree trunks and rocks. Thus, the prevailing shade of the environment in industrial areas has changed from light to dark. In these areas some seventy species of moths, which have protectively colored wings, have also changed from light-winged forms in the mid-nineteenth century to the dark forms of the present. In the days when the light forms of the species were prevalent, collectors occasionally found black mutations. Under changing conditions natural selection, acting mainly through predation by birds, favored the black mutants so that they gradually gained ascendancy and are now common. Experimental work shows that under field conditions the black form now has the advantage near industrial areas when it comes to escaping the attentions of birds.

A few moths fly in the daytime. It is interesting that these are often brighter than their nocturnal relatives, which are of somber pastel shades or are concealingly colored. The tropical and diurnal uraniids are as brilliant as any tropical butterfly. The day-flying moths common in North America include conspicuously marked mimics of wasps and bees.

Night-flying moths are able to find flowers in near darkness, presumably both by scent and sight. Some deep purple flowers, invisible in dim light to the human eye, are specialized for nocturnal pollinators and are believed to reflect ultraviolet light to a degree that enhances their visibility to insects. It may be that the positive reaction of moths to light is related to their flower-visiting habits. Apparently strong light repels them, but, before they can sheer away from a very bright light, they become helplessly dazzled and are unable to escape.

The most ancient fossil Lepidoptera known are no older than early Tertiary and differ little from modern types. The structure of living Lepidoptera and Trichoptera shows, however, that the butterflies and moths were derived either from Trichoptera or Trichoptera-like insects. Lepidoptera, then, are probably descended from aquatic ancestors, but their relationship to the primitive aquatics is not understood.

The Lepidoptera transformed the hairy covering on the wings of the Trichoptera into a dense covering of scales and also greatly increased the general hairiness of the body. It is interesting to speculate on the functional significance of the scaly wings of the Lepidoptera. The theory most often cited is that the scales serve to strengthen the wings, which have comparatively few cross veins. This theory is probably not correct. There are some butterflies with naked wings, without any changes in venation to compensate for any weakness that might result.

Another theory, which may be correct, is more involved, but it is based on the idea that the hairy covering of moths (Fig. 60), including the scaly covering on the wings, is an insulating blanket to conserve body heat. After a moth has been in flight for some time, its body becomes quite warm as a result of the heat generated by the flight muscles. A large hawkmoth caught on the wing feels as warm as a bird. The wing muscles are thus necessarily designed to operate at relatively high temperatures, and it can be observed that moths have difficulty flying when their muscles are cold. On a chilly evening, a large moth that has been at rest can not take off immediately but has to quiver its wings for a time before it can fly. If the body is well insulated, the blood and the "motor" can be warmed up more quickly than otherwise. Since the moths are active mostly at night, when temperatures are lower and when there is no chance of getting warmed by sunlight, insulation would be quite important. Since blood circulates through the wings, perhaps the scaly covering there also is essential to prevent heat loss.

Butterflies are undoubtedly derived from moths, and here the function of the wing scales has perhaps been altered. The body of butterflies is not so well insulated as that of moths, and their mode of life would make insulation less valuable. Perhaps the scales on the wings have been retained, even though no longer needed for insulation, because they had taken on the function of providing the color pattern of the wing, which is undoubtedly of great significance— primarily for sexual recognition—in the life of the butterfly.

Caddis flies (order Trichoptera) are moth-like insects that usually fly at dusk. The group is a small one compared with the Lepidoptera, there being only about five thousand described species. They have longer antennae than moths, are usually without color patterns on the wings, and are usually more agile afoot than moths, leaping into the air to take wing, and running again after alighting (Fig. 62). The adults have mouthparts apparently adapted for lapping. The mandibles are weak or vestigial.

Larvae of the Trichoptera are aquatic. Some spin underwater "spider webs" to catch drifting prey, webs that are strong enough to catch newly hatched fishes, and some of these caddis flies are pests in fish hatcheries. The best known caddis flies build portable cases of

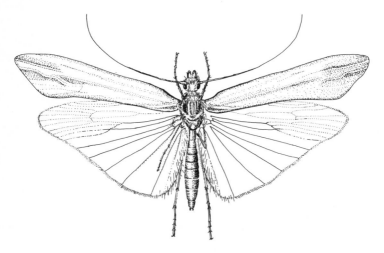

FIG. 62. *Caddis fly.*

sand, small empty snail shells, or bits of vegetation bound together with silk. The shelter is highly characteristic of the species of caddis. Some are neatly coiled, like a snail shell (Fig. 63), and in fact a caddis-fly case once was described as a new species of snail. Others, found especially in stagnant water, are built of grass stems stacked like the logs of a log cabin (Fig. 64). If larvae of the sand-case type are carefully removed from their shelter, then supplied with small glass beads, they will build a new case with the beads. One entomologist found in a small Nevada stream caddis-fly cases studded with tiny fire opals.

FIG. 63. *The examples below of insect architecture are protective, portable cases constructed from sand grains by the aquatic larvae of caddis flies. These cases, each about one-fourth inch in diameter, have been exposed to weathering, and the spiral interiors of some have been exposed.*

The opals were more abundant in the sand of the cases than in the sand of the stream bottom.

Both in the adult and larval stages, caddis flies are important in the dietary of fishes.

FIG. 64. *This portable shelter was constructed by a caddis-fly larva from bits of dead vegetation. It is about one-half inch long.*

19

The two-winged flies

The Diptera, or two-winged flies, are fragile, soft-bodied creatures; yet nearly all of the fiercely-biting insects are Diptera, and great is the annoyance, pain, and number of deaths that they inflict on human beings.

In all Diptera the hind wings are missing or, rather, have been modified into a pair of sense organs, or halteres. Other insects that use only the front wings in flight are some of the mayflies, the males of scale insects, and the strepsipterans.

During flight these sense organs vibrate rapidly, each in a single plane. A haltere consists of a relatively heavy knob on the end of a slender stalk (Fig. 65), and such a vibrating system resists any change in its plane of movement, just as the whirling, heavy wheel of a gyroscopic resists a change in its plane of movement. The organ is thus called an oscillating "gyroscope" and functions to detect changes in the orientation of the fly. Should the fly shift its orientation, the resistance offered by the vibrating halteres is translated into a shearing force at their bases, which is detected by sense cells situated there.

The larva of the fly always differs greatly in its mode of life from the adult. The highly specialized larvae of Diptera lack all traces of legs, and some are without recognizable heads (Fig. 66) and seem to be little more than digestive tracts swathed in layers of fat that accumulate as a result of incessant feeding by the larva.

When it transforms into the pupa, the fly larva does not spin a cocoon, but, in the more advanced Diptera, the last larval skin is not shed but becomes thickened and is retained as a shelter (the puparium) for the delicate pupa. The soft, jawless adult fly gets out of the puparium by means of a special organ, the ptilinum, which is a sack pushed out of the head through a crescentric slit. Filled with fluid under pressure, this sack forces open a circular lid at the end of the puparium.

Several groups of quite unrelated flies have independently become specialized for drinking blood of vertebrate animals. Because of their numbers, their pertinacity, and the fact that they transmit disease, they have profoundly influenced the numbers and distribution of vertebrate

FIG. 65. *Visible as a light-colored structure near the center of the photo is the haltere, or balancing organ, of a tachinid fly. The white shields above the haltere and below the wing are the calypteres, of unknown function.*

animals, including man. Among the more primitive of the blood-sucking flies are the mosquitoes (family Culicidae), flies with active, aquatic larvae.

The male mosquito feeds on nectar; females of some species also are nectar feeders, but most have to feed on blood to produce offspring. It is surprising that two very closely related species of mosquitoes, hardly distinguishable in appearance, may differ strongly in their biologies, one requiring a blood meal, the other able to produce mature eggs from protein reserves accumulated during larval life.

The slender beak of the mosquito looks superficially like the beak of the Hemiptera but differs in detail. Both mandibles and maxillae

FIG. 66. *Larva of an advanced two-winged fly.*

are long, piercing stylets. The drinking tube is not formed from the paired appendages but from two unpaired structures: the labrum, or front lip, and the hypopharynx, a tongue-like structure, concealed in most insects, that lies against the inner wall of the labium. The labium is a relatively stout sheath that houses the hair-thin piercing beak.

Most mosquito larvae feed on microorganisms, which they sweep out of the water with mouthparts shaped into brushes. A few of the larger species are predators, capturing other mosquito larvae. Although the mosquito larva is legless, it can swim by movements of the whole body, hence the name wriggler. It spends most of the time near the surface, with the caudal breathing tube opening out into the air. *Anopheles* lies horizontal, with the mouth brushes at the surface film; *Culex* and others hang head downward. When disturbed, the wriggler closes up the breathing tube and dives. The pupa also is aquatic, also can swim actively, and breathes at the surface through a pair of breathing tubes on the thorax. Wrigglers live mostly in shallow, temporary bodies of water that do not contain fish, and an effective method for controlling mosquitoes breeding in sizable ponds is to introduce small fish. The mosquito is able to find very small aquatic habitats: the cut, water-filled stem of a bamboo may house several larvae; and a plague of mosquitoes in a department store was traced to the water wells in gummed-tape dispensers.

Malaria and yellow fever, both transmitted only by mosquitoes, are important in human ecology, and it has been said that they have been the direct or indirect cause of about half of human deaths. After the discovery, in the 1880s, that mosquitoes were the vectors of *Plasmodium*, the malaria parasite, the classification of mosquitoes was studied intensively. Linnaeus described two species of mosquitoes; by 1905, 450 species were known; the number jumped to 1,050 in 1910, rose to 1,400 by 1932, and new species are currently being described at the rate of about 50 a year.

The systematist usually maps a fairly extensive terrain and considers his area of competence to include hundreds or a few thousands of species. He does as well as he can with the information at his disposal —mostly the anatomy of dead museum specimens. It is easy to say that he should also know the biology of the species, but in fact, if he took the time to get this information, the area he could map even in a lifetime would be far too small to be of any use to guide the specialists to the limited problems they work out in detail.

When the scientists working on malaria concentrated on the few species of mosquitoes relevant to their problem, they found that existing classifications amount to little more than sketch maps. For example, the important malaria mosquito of Europe, *Anopheles maculipennis*, proved to be a complex entity. One sample of *maculipennis* preferred to bite human beings, but another preferred domestic animals. One sample carried malaria, another did not. Mating behavior varied from one lot to another. As the tangle was gradually sorted out, it became apparent that the species *Anopheles maculipennis* was actually a complex of several species. These species, nearly identical so far as structure of the adult was concerned, lived side by side, but were kept separate in nature by differences in mating behavior and choice of habitat. Renewed study of the morphology of all stages showed that the eggs of the hidden, or "cryptic," species were distinctive, and could be used to identify them.

Fungus gnats (family Mycetophilidae) are delicate, mosquito-like, nonbiting flies. The larvae of most live in mushrooms, a minority are carnivores. In the cave at Waitomo, in New Zealand, is a great colony of mycetophilids, whose larvae, which glow like fireflies, cling to the roof of the cavern and sprinkle it with golden light. They spin sticky threads to trap midges that grow in the waters of the cave. When a midge flies toward the glowing larva, attracted by the light, it strikes the thread and is hauled up to be eaten. Related but non-luminous species in other parts of the world spin their webs over damp logs and under stones, beading them with poisonous droplets of oxalic acid that kill on contact.

There are other groups of primitive flies that feed on blood. Buffalo gnats, or black flies (Simuliidae), bite fiercely and venomously. Immense swarms sometimes appear and have been known to kill cattle and horses in a few hours. The venom injected is an allergen that renders some individuals highly sensitive to the bite of these flies. Moth flies (Psychodidae), so named from the densely hairy wings, include both biting and nonbiting species. Species of *Phlebotomus*, which live in warm regions, transmit the fatal Oroya fever of South America and sandfly fever. Punkies are gnats (Ceratopogonidae) so small that they go through mosquito netting with wings outspread. Their bite is nevertheless a painful one.

Among the flies that often are confused with mosquitoes are the crane flies (Tipulidae). None of the six thousand or more species of this family can bite, but some of the large kinds, with a wing span of over two inches, are the basis for stories of gigantic mosquitoes. Midges (Chironomidae) are closely related to both mosquitoes and punkies. None are biters. They exist in very large numbers, and both the larvae and the adult midges are an important item of the food web of natural communities.

All of the primitive biting flies are small, weakly flying insects, but the more advanced types tend to be large and active. In the family Tabanidae, the deer flies and horse flies, are some of the largest Diptera, and all are swift fliers. Tabanids rarely get their fill of human blood because their bite, about as painful as a puncture with a large-gauge hypodermic needle, does not often go unnoticed. The deer flies, with spotted wings and rainbow-iridescent eyes, are annoying pests in Eastern woodlands. That they are specialized for preying on animals that are more furry than human beings is shown by the fact that they dive persistently into one's hair.

The belief that houseflies bite at certain times of the year is based on the fact that the stable fly (*Stomoxys*), a fierce biter, closely resembles and is in fact a close relative of the housefly (family *Muscidae*). The tip of the labellum, which is the soft sponge of the housefly proboscis, is here furnished with sharp teeth that rasp through the skin.

Also closely related to the housefly are the tsetse flies (*Glossina*), which feed on the blood of large mammals. In mid-Tertiary, much of North America probably resembled the African savannah country and teemed with large spectacular mammals. Tsetse flies also were here at that time, for their fossils have been found in the Florissant shales of Colorado, but they have since become extinct everywhere except on the African continent. They probably are the most intensively studied insects in the world, for they carry the lingering and fatal sleeping sickness that has killed Africans by the hundreds of thousands.

The larval tsetse fly lives in the uterus of the parent, feeding on the secretion of "milk" glands, until full grown. The parent deposits the mature larva on the ground, where it burrows from sight and pupates. Only one is born at a time, and the parent requires a full blood-meal for each of the young.

Among the several groups of flies that, in the winged stage, prey on other insects, the robber flies (family Asilidae) are the most varied and powerful (Fig. 67). The robber fly usually takes up a post on a bare patch of ground, or on a stem that commands a view for some yards about, and darts out after passing insects. The larger kinds overpower grasshoppers, dragonflies, bumblebees, and honeybees, and even other asilids. After penetrating the armor of the victim with the short beak, the robber fly drains out the body fluids, perhaps predigesting the tissues. The legless larvae of asilids also are predaceous, prowling slowly through earth or decaying wood after soft-bodied insects.

Certain small, obscure flies of the family Milichiidae habitually ride on the backs of robber flies and join their host in its meal when a

FIG. 67. *A predaceous two-winged fly of the family Asilidae.*

capture is made. Most of the species of this family feed on insects freshly caught in spider webs.

The very small, brilliantly green flies that are often seen running about over leaves are the long-legged flies (Dolichopodidae). They pounce upon small insects that are encountered.

Another group of delicate predatory flies are the dance flies (Empididae), named for their habit of gathering for a slow up-and-down aerial dance. In certain species, the male carries a shining balloon, made of some secretion, which is presented to the female in courtship. There is a wide range of variation in mating behavior in this extensive family of flies, which perhaps makes it possible to trace the evolution of the balloon procedure. Apparently the most primitive example of the present-making technique is found in species that present the female with a captured insect. It is suggested that this behavior may at first have made it safer for the male to approach the female, as she normally preys upon insects about his size. Other species wrap the offering in a glistening web spun from glands on the front legs. Yet others inflate the package into a conspicuous, frothy balloon; this is presented to the female, who devours the enclosed insect. A more advanced stage is exhibited by species that may enclose only a small insect, which is not eaten, or even an inedible bit of plant material. This could lead finally to the completely aesthetic situation in which the balloon alone is given.

One of the great groups of insects which parasitize other insects is a family of some five thousand species of generally dingy, bristly flies, the Tachinidae. The larvae of the tachinids feed inside their host and are usually large enough or numerous enough to kill it. The victim usually is a caterpillar. The female may glue the eggs or new-born larvae onto the host or may simply scatter them in the area likely to be visited by it. One species leaves its young on the silken trail that the host caterpillar makes between its retreat and feeding place. Some strew thousands of minute, hard-shelled eggs on leaves likely to be eaten by the host; the eggs hatch inside its digestive tract. Some tachinids have a thorn-like structure at the end of the abdomen that is pressed through the skin of the host to get the eggs inside. The tachinids are abundant and in nature exert a strong influence on the insect fauna. Several species have been introduced into the United States to combat such harmful immigrants as the gypsy moth and Japanese beetle.

Some dipteran analogues of vultures are the blow flies (Calliphoridae), mostly brilliant blue or green flies that lay their eggs on dead animals. The larvae consume the flesh, first liquefying it by flooding it with digestive enzymes. So certainly do these flies find a carcass that the ancients thought decaying fish changed spontaneously to maggots. A fish thrown on the bank likely will in an hour or two have its gills stuffed with masses of the white blow-fly eggs. Blow-fly maggots excrete allantoin, urea, and other substances that inhibit the growth of bacteria. Doctors who dealt with battlefield wounds in World War I observed that old wounds infested with maggots were less likely to become gangrenous than those that they had cared for promptly. This led to the use of sterile maggots in surgery, especially in the treatment of bone infection, but they have been replaced by modern antibiotics. The sheep fly (*Lucilia*) commonly scavenges on dead animals but has learned to lay its eggs in the soiled, heavy wool of living sheep. The maggots bore through the skin, sometimes in numbers great enough to kill the animal. Some calliphorid maggots live in bird nests, drinking the blood of the nestlings. The Congo floor maggot has somewhat similar habits; it lives under sleeping mats and comes out at night to feed on the blood of human beings.

The Bombyliidae, or bee flies, delicate fuzzy insects with long beaks used for drinking nectar, are in their larval stages parasites of ground-nesting insects. The newly hatched larva is a minute, wiry creature that actively works its way through the soil in search of its food. The larva that preys on bees samples buried honey and pollen until its victim becomes full grown, then transforms into a sedentary grub that sucks out the contents of the bee larva. Neither the grub nor the fluffy adult fly would be able to get out of the ground into the open air. This is accomplished by the active, hard-shelled pupa, which is equipped with spines and blades that enable it to cut, pry, and push its way out. Like the tachinids, the bombyliids are abundant, and those species that live in grasshopper egg-pods are credited with checking the numbers of some potentially harmful grasshoppers.

Most of the many thousands of species of so-called parasites that, like the tachinids, kill their host are not really parasites in the classical or medical sense. Some of the Diptera, however, are "true" parasites. The larvae of the flies once placed in the single family Oestridae, but now divided into four separate families, all live inside mammals, hosts which are much larger than the parasite and which consequently are usually not killed by it.

The larvae of the horse botfly (*Gastrophilus intestinalis*), one of several similar species in the family Gastrophilidae, cling in dense masses to the wall of the stomach of their host, where they absorb nutrient from both the stomach wall and its contents. The female fly darts in repeatedly to lay her eggs, one at a time, on the hairs of the animal. When the horse licks its hide, the newly hatched, slender, spiny larva burrows into the skin of the tongue, then in later stages moves to the stomach. The mature grub leaves the host with the feces. The fat of the young larva is reddened by hemoglobin, apparently used to store oxygen.

The warble fly of cattle, *Hypoderma lineata*, can be taken as an example of the family Hypodermidae, whose larvae grow in the body spaces between the internal organs and the muscles. When the female warble fly lays her eggs on the cattle, the animals become panic-stricken—some of the warble flies are called bomb flies—and flee in terror, although the egg-laying process is painless. The eggs, laid on hairs, soon hatch, and the larvae bore in for a period of several months of feeding and wandering over the viscera of the host. As they near maturity, they gather under the skin of the back. Here the tumor-like swellings, or "warbles," are formed; the larvae remain in these during the winter. The full-grown, spiny, inch-long maggots emerge from the warbles and fall to the ground, where they pupate. The perforated hide is worthless for leather, and the adjacent meat—the area from which the best steaks are taken—is made unfit for sale.

Parasites that specialize in infesting the nasal cavities of the host are the head bots, of the family Oestridae (in the limited sense). They attack a variety of mammals, including such dissimilar hosts as marsupials and man. The sheep botfly (*Oestrus ovis*) is viviparous and deposits its minute young in the nasal openings of its host. As with the warble flies, the painless attack of the female stirs the victim to a frenzy. The larvae complete their development in the nasal sinuses and reach a length of over an inch. Not infrequently the host is killed by the infestation. This fly sometimes darts into the eyes of human beings to deposit its larvae, and it is said that shepherds in North Africa have been blinded by the spiny larvae.

Most species of the large flies in the family Cuterebridae parasitize rodents, but one, *Dermatobia hominis*, infests man as well as several other animals. For some yet unfathomed reason, the female does not lay her eggs directly on the host. Rather, she lurks about the host and lays her eggs on some other species of fly in the neighborhood, usually a mosquito. When the carrier alights on the host, the egg hatches quickly, and the larva burrows into the skin. It completes its development in the subcutaneous tissues.

One group of Diptera—the Hippoboscidae and related families— have become intensively specialized for life as blood-drinking ecto- parasites on birds and mammals. They cling to their host, have the body flattened and tick-like, and some, such as the common sheep tick (*Melophagus ovinus*), are wingless. These remarkable insects carry the larva in the uterus until it is full grown and ready to pupate; thus, the adult represents the only feeding stage in the life cycle.

The family Muscidae, one of numerous species of dull-colored, scavenging flies, does not readily fit into the preceding categories and would escape notice but for the fact that one of its species, *Musca domestica*, is common in cities and towns. The housefly was especially abundant in cities when horse dung—a favorite habitat of the larva— was available in quantity.

When DDT was developed in the 1940s, the housefly responded in a way that was of considerable theoretical interest to the biologist. All animals vary, not only in their external anatomy, but also in the finest details of internal biochemical functioning, even to the detailed structure and function of the enzymes. Apparently the normal house- fly has enzymes that disassemble natural organic molecules related to DDT, and included in the variants of such enzymes are some able to take apart the insecticide molecules and render them harmless. The very heavy mortality, over wide areas, that was inflicted by DDT created enormous selection pressure favoring the variants, and, in the course of a few years, local populations of houseflies have evolved that are difficult to kill with DDT. It is quite possible that reproductive barriers between these resistant populations and the old-fashioned houseflies will gradually evolve, thus creating new species.

In the family Drosophilidae are several hundred species of minute flies that are little noticed except for the species *Drosophila melanogaster* that hovers about overripe fruits in kitchens and markets. The larvae of this fly live in the decaying fruit where they feed on the micro-organisms, particularly yeasts, that flourish there.

About the time that Mendelian inheritance was rediscovered (1900), the entomologist C. W. Woodworth suggested to a group of biologists working on the effects of inbreeding that they might well use the small fly *Drosophila melanogaster*, since it takes up little space, and generations follow each other quickly, as often as every ten days. A few years later the fly was given to T. H. Morgan to demonstrate in the class-room the inheritance of eye color. In the hands of Morgan and his collaborators, this fly subsequently provided most of the information used in erecting the science of modern genetics. By studying the results of crosses between flies that differed in various ways, the concept was built up of units of inheritance (genes) that were strung like beads on the chromosomes. It is an astonishing coincidence that, long after the existence of these strings of mnemonic beads had been demonstrated

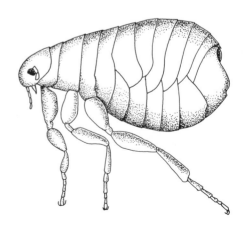

FIG. 68. *Flea.*

by indirect evidence, *Drosophila* turned out to be one of the few animals that had giant chromosomes, hundreds of times larger than ordinary chromosomes. These occur in the cells of the salivary glands of the larva. In them the structure observed under the microscope corresponds to many of the predictions that had been made on the basis of breeding experiments.

Most orders of insects seem to have near relatives, but the fleas (order Siphonaptera) are like nothing else. Because their larvae are legless maggots, the fleas usually are listed next to the Diptera, but there probably is no real evolutionary relationship between them.

None of the five hundred or so species of fleas have a trace of wings (Fig. 68). All are, as adults, blood-drinking external parasites of warm-blooded animals. Unlike other insects, they are flattened from side to side; this body form makes it possible for them to slither quickly through the hairs or feathers of their host. Their extreme hardness—they feel like a flat seed—makes them nearly immune to fingernails or teeth of their tormented victims.

The blood-drinking habits of the fleas make them vectors of animal and human diseases. The Black Death, or bubonic plague, is caused by a bacterium transmitted by several species of fleas. Some of these species feed indiscriminantly on a variety of animals, including man, so that although the plague is primarily a disease of rodents, under proper conditions it is transferred to and becomes established in crowded human populations, where it has caused immense mortality. In the United States, the disease smolders in certain populations of wild rodents.

20

On stings and societies

The immense order Hymenoptera is one of insects with hard bodies, tough, glassy wings, and sharp stings. When one tries to summarize their biology, one thinks of the intense vitality of these insects: of ants frantically tumbling over obstacles as they drag booty to their nests; of wasps running over the ground in nervous haste in search of their prey; of bees dashing from flower to flower gathering pollen with movements too quick for the eye to follow. Such a characterization is not entirely inappropriate, since this activity is directed toward the welfare of the young, and it is to the good start in life given the young that this order of insects owes much of its success.

In some Hymenoptera the care of the young has reached a degree of refinement in which there is division of labor between individuals that produce young and those that care for them. This differentiation between reproductive and nonreproductive castes forms the basis for insect societies, and these reach their greatest variety and complexity in the order Hymenoptera.

A structure that has been of strategic importance in the history of this order is the ovipositor. In the most primitive Hymenoptera, it is used to force eggs into the shelter of plant tissues. It next is used to insert eggs into other insects, a development that laid the foundation for the evolution of an immense array of parasitic types. In a subsequent evolutionary step, it is used as a hypodermic needle for injecting an anesthetic that makes it possible to store live food in quantity for the young. And, finally, it becomes a fiery sting used solely for defense.

The four naked, transparent wings of the Hymenoptera are of unequal size, and the smaller hind wing fits against the front when outspread in such a way that the outline of both together is that of a single wing. Moreover, the hind wing grasps the front with a row of strong hooks, so that the two can be pulled apart only with difficulty. The resulting single pair of flight surfaces is vibrated rapidly in flight, with a frequency as high as two hundred and fifty times a second in some of the more powerful fliers. This type of wing action confers on the Hymenoptera the possibility of very precise control. The insect can hover stationary in the air, instantly dart forward, pause again, and swerve quickly from side to side. But not all are good fliers; some of the most successful Hymenoptera are wingless terrestrial insects.

The feeding appendages of the adult are of a somewhat specialized type. All are equipped with biting mandibles that, although of rather ordinary appearance, are often more important "industrially" rather than nutritionally, since they may be used for nest building. The maxillae and labium are fused together to form a lapping structure adapted for liquid food.

In all but the most primitive Hymenoptera the thorax is unique, being constructed of four rather than three segments. This type of thorax is correlated with a "wasp waist," a constriction between the first and second segment of the abdomen. A narrowing at this point necessarily adds a segment to the thorax, subtracts one from the abdomen. The waist is a narrow hinge-joint that confers great mobility and makes it possible to wield the sting effectively.

In the Hymenoptera, the sex of the individual depends on whether or not the egg from which it developed was fertilized. Unfertilized eggs usually develop into males, and the male therefore has only a haploid complement of chromosomes.

Larvae of the Hymenoptera range from moderately active caterpillar-like types in the primitive groups to legless, maggot-like larvae. These helpless grubs often live in a cell dug in earth or wood and stocked with food by the parent. Usually such larvae have an incomplete digestive tract, in which the hindgut has not established a connection with the midgut. The larva then does not defecate until it matures; this apparently helps to keep the food supply from being contaminated with molds and other microorganisms, which are some of the most serious enemies of these Hymenoptera.

Corresponding to the difference in thoracic structure, the Hymenoptera are divided into two suborders: the thick-waisted group (Suborder Chalastogastra), and a wasp-waisted one (Clistogastra). The first suborder is much the smaller with regard to number of species, is the more primitive, and is one with vegetarian larvae. Larvae of the suborder Clistogastra mostly are carnivores, although there are many that live in galls or eat pollen.

Judging from their structure, the plant-eating Chalastogastra are the more primitive of the two main groups of Hymenoptera, since the thorax is of the usual insectan type (three segments), the wing venation is relatively complex, and the larvae are not as degenerate as those of other Hymenoptera. The oldest known fossil Hymenoptera, from the Upper Jurassic, belong to this primitive suborder, the wasp-waisted Hymenoptera not appearing in the geological record until Cretaceous time.

The most successful surviving group of the primitive suborder is that of the sawflies (Tenthredinidae), of which there are about three thousand species, some of them abundant enough to be destructive to forests. They have as part of the ovipositor a pair of saw-toothed blades used to make incisions in plant tissues to shelter the eggs. To some extent the serration on the blades is correlated with the kind of material that has to be sawed—coarse teeth when the eggs are laid in woody twigs, fine teeth for soft green tissues.

Larvae of the sawflies closely resemble the larvae of the Lepidoptera. They have prolegs to support the abdomen, which, however, adhere by suction and are not equipped with the hooklets found on the prolegs of caterpillars. Many species defend themselves with malodorous secretions, and these often advertise themselves by being brightly colored and gregarious. When threatened by their hereditary enemies, the ichneumon flies, they thrash the abdomen about violently to ward off the attack. The chemical defenses perhaps are directed more particularly against birds. The green, concealingly colored larvae are sometimes rendered even more nearly invisible by the long hairs over the body, as these keep it from casting a sharply outlined shadow.

Perhaps the Hymenoptera would not have achieved any great success had they remained plant-eaters. The suborder with this mode of life is much the smaller one and is far inferior in number of species to the order Lepidoptera, a group with a more or less equivalent way of life. Those groups that took up the practice of feeding, during the larval stage, on other insects founded the groups of Hymenoptera that now make the order a large and important one.

There are living today some rare and aberrant representatives of the suborder Chalastogastra whose behavior may indicate one of the ways in which the transition from plant-eating to insect-eating took place. These are the oryssid wasps, which with their stout, sharp ovipositor thrust their eggs into wood. The larvae, unlike those of their near relatives, feed on wood-boring insects, not on wood. Once such insects became specialized for placing their eggs near or eventually on their insect host, the basis for the kind of parasitism found in the parasitic Hymenoptera would be established. It is quite possible that the transition from a herbivorous to carnivorous mode of life took place in more than one group of primitive Hymenoptera and proceeded in different ways.

The parasitic Hymenoptera, termed the "Parasitica," are distinguished from the true wasps by the fact that the ovipositor issues from the underside of the abdomen and some distance before the tip. In the true wasps (the "Aculeata") the ovipositor or sting is terminal. There are four major groups of parasitic wasps: the ichneumons, the chalcids, the gall wasps (of which only a minority are parasites), and the serphoids. These groups usually are ranked as superfamilies.

Two of the more important families of parasitic wasps are the Braconidae and Ichneumonidae, two related families of small to large insects whose larvae devour spiders or other insects.

Braconids are on the average smaller than the true ichneumons. Some of the common ones fit comfortably inside plant lice, a single individual requiring the contents of only one plant louse to reach maturity. Other species live in crowds on or inside caterpillars. One often finds tight clusters of tiny cocoons of the caterpillar-infesting braconids on dead plant stems. Such a cluster contains the offspring of one female wasp. She finds a good-sized caterpillar, mounts it, and, while it thrashes about, pierces its skin with her ovipositor to squirt the eggs inside. The braconid larvae at first feed on blood and fat tissues, leaving vital organs intact, so that the caterpillar continues to live and feed, converting plant into animal substance for the benefit of the parasites inside it. Toward the end of their larval life they damage the host lethally. At about the time the caterpillar would normally have become full-grown, the white, legless, maggot-like larvae bore out through the skin and, crowded together a short distance away from the dying host, spin a mass of tough, yellowish silken cocoons. Other braconids parasitize ants and the larvae of beetles and of Diptera. Very few parasitize other parasitic Hymenoptera, a practice (hyperparasitism) that is common among the Ichneumonidae.

The ten thousand or more species of ichneumonids are parasitic on other insects or spiders. Collectively they must know a great deal about the insect world, for they successfully hunt out an enormous number of species in exceedingly diverse places. Some specialize on a single kind of prey, others attack a wide range of species scattered through several orders. They find their prey in turf, in the foliage of shrubs and trees, and in a variety of protective shelters; they can sense the presence of their victims beneath wood and even dive into the water after their prey. The life histories of the species are unendingly varied. There probably are more species of ichneumons yet to be named and described than are already known.

Larvae of many ichneumons cling to the skin of their host and feed through puncture wounds. The parent of such larvae may paralyze the host with its sting, in the manner of the true wasps. A spider parasite (*Polysphinctus*) stings and paralyzes its host, then glues on it a single egg. The spider soon recovers and is apparently normal, except that the wasp venom has somehow rendered it unable to molt. The egg, and later the grub of the parasite, is thus not in danger of being cast off, as it would be were the spider to shed its skin in the normal fashion. While the spider is going about its business of catching insects and sucking out their body contents, the parasite has delicately opened the abdomen of the spider and slowly feeds on its living tissues. Near the end, before it is finally killed by the parasite, the spider is carrying on its back a fat grub half as large as itself.

Larvae of other species live inside their host, sometimes in such numbers as to nearly fill it, and pack the interior of the empty skeleton with cocoons at maturity. Those that oviposit inside the heavy cocoons of moths may be attracted in swarms by the odor of freshly spun silk, and the females may insert hundreds of eggs into the caterpillar. When the host is thus "superparasitized," the number of parasite larvae is reduced by their eating each other.

The spectacular North American ichneumon *Megarhyssa* uses its four-inch-long ovipositor to reach the larvae of large wood-boring Hymenoptera. In some way the female is able to determine the location of the borer. She can drive the slender, flexible ovipositor in a matter of half an hour or so through an inch or more of solid wood to reach the host. Sometimes several of these large ichneumon wasps will be seen in a group over a promising site. They stay put, being unable to withdraw the ovipositor quickly.

The feeding habits of such parasites of bees as *Grotea* and its relatives illustrate the nutritional equivalence of meat and pollen. The larvae of these ichneumons first fill and eat the egg or young larva of the host bee, then complete their growth on the store of pollen and nectar that was laid up by the bee. They are the only herbivorous ichneumons.

Hyperparasites prefer the flesh of other parasites to that of the larger host insect. Thus, the female of the ichneumon *Hemiteles* attacks the caterpillar of the cabbage butterfly, but her larvae feed on larvae of the braconid parasite *Apanteles* that live inside the caterpillar. An occasional third member of this intimate community is the larva of a chalcid wasp that feeds on *Hemiteles*. If we were to rank these insects according to whether beneficial or injurious, *Apanteles* would be good, *Hemiteles* harmful, and the tertiary parasite, the chalcid, again a beneficial one.

The chalcid wasps are the largest group of parasitic Hymenoptera. There are many thousands of species yet to be described, and the intricate and diverse life histories of most remain to be worked out. Conservative classifications combine them into a single family, others divide them into twenty or more families. In general they are small or minute, some so tiny that thirty wasps may emerge from a single parasitized insect egg.

Although most chalcids are parasites of other insects, many are herbivores, a difference that does not seem very fundamental, since even rather closely related species may differ in this way. In the subfamily Eurytominae, for example, there are both parasites and vegetarians. The larvae of the plant-eaters often live inside seeds or in stems, where they may induce a gall. Those of this subfamily that are parasites feed on other insects that live in such places.

The chalcid wasps that make galls in the flowers of figs cross pollinate these plants. Since some of the cultivated varieties, as the Smyrna fig, will set fruit only if cross pollinated with another variety, the wasp must be present if the crop is to be grown. The introduction of the Smyrna fig into California was unsuccessful until followed by the introduction of the proper wasp, a species of *Blastophaga*. The male of this chalcid is wingless and does not leave the fig but runs about on the flowers that line the fruit until he finds a gall that houses a female and cuts through it to mate. The winged female then leaves the fig to find fresh flowers in which to oviposit.

A gall-making chalcid that is abundant enough to be important to the farmer is the wheat jointworm (*Harmolita tritici*), a native insect that lives in grass but also infests the stems of wheat.

Among the parasitic chalcid wasps, the parent as well as the young may feed on the host insect. When the insect that is parasitized lives in places out of easy reach, the wasp may use her ovipositor both to get the eggs into the host and to obtain a drink of insect blood for herself. A species of *Habrocytus*, for example, parasitizes a small caterpillar that hollows out the inside of dried grains of corn. When the caterpillar is grown, it prepares for pupation and for the emergence of the moth by cutting an exit hole, then sealing it over with a thin sheet of silk. When the chalcid finds one of these silken membranes on a kernel of grain, she drives her long ovipositor through it to reach the caterpillar and pricks the victim so that blood oozes out. She then sets about making a drinking tube. A drop of liquid is exuded from the ovipositor onto the underside of the barrier. By working the ovipositor downward, this material, which quickly hardens, is extended into a tube. Adding more material, the wasp builds the tube down to make contact with the wounded caterpillar. She then withdraws the ovipositor and turns around to drink the blood that wells up.

One chalcid (*Eurytoma monemae*, of China) successfully parasitizes a caterpillar that is sheltered by a cocoon too tough for it to penetrate by using the services of a larger and stronger parasite, one of the cuckoo wasps (*Chrysis*). The female chalcid is attracted, not by the cocoon of the caterpillar, but by the chrysid wasp, whom she follows about until the chrysid, itself a parasite of the same caterpillar, finds the cocoon. The wasp tears open the cocoon to lay her egg on the caterpillar, after which she loosely plugs the opening. This plug can be pierced by the ovipositor of the chalcid, who proceeds to lay a number of eggs. Her larvae destroy the chrysid young and devour the caterpillar or, sometimes, live as secondary parasites on older chrysid larvae.

The small chalcid wasps usually parasitize rather small insects—scale insects are favorite hosts, for example—or insect eggs. An adaptation for utilizing large hosts efficiently is found in the subfamily Encyrtinae. This adaptation is polyembryony, in which a single egg, after insertion in the host, divides and subdivides to produce a large number of eggs, sometimes as many as a few thousand. Each egg then develops into an embryo and eventually a wasp.

In the gall wasps the situation that exists in the chalcid wasps is reversed, for the vegetarians are in the majority. About three fourths of the two thousand species are gall-makers. Some species are inquilines, living in galls that were produced by other gall wasps. Most of the gall-makers infest the single plant genus *Quercus*, that of the oaks. The swellings they make on the twigs and leaves of oaks may be as large as apples and house hundreds of the tiny insects.

During the winter months the larvae of the oak-gall wasps live in galls on the roots of the tree. All develop into wingless females, which emerge from the ground in the spring and climb up the tree to lay eggs on the young twigs. Reproduction is parthenogenetic, there being no males in this wingless generation. As the oak pours nutrient into the twig, most is diverted to form the tumorous gall that surrounds

FIG. 69. *The powerful sting of the cicada-killer wasp issues from the tip of the abdomen, as in all bees and the true wasps. Inside the abdomen of these insects are venom glands whose secretion pours out through the sting.*

the wasp larvae. After feeding throughout the summer, the larvae pupate inside the gall. The winged mature wasps, which in this summer generation include both males and females, chew their way out of the gall. After mating the female burrows into the ground to lay her eggs in the roots.

An oak gall may house a varied community of insects that eat either the substance of the gall or the other insects; some thirty species of such insects, belonging to several orders, have been known to emerge from the galls that were grown by one species of gall wasp.

The parasitic Hymenoptera that belong to the superfamily Serphoidea bridge the gap between the Parasitica and the Aculeata (true wasps). The sting is terminal, but in other ways they resemble the parasitic wasps already discussed. Most of the species are minute parasites of the eggs of other insects.

Many species of the family Scelionidae use other insects for transportation, a practice called phoresy. The female wasp clings to the female of the insect whose eggs she intends to parasitize. The wasp *Rielia*, for example, whose young live inside the eggs of the praying mantids, rides her host until the mantid deposits eggs. She then slips off and inserts her eggs into those of the host. The mantid proceeds to elaborate the protective egg cover of hardened spongy material, but too late to ward off the attack of the parasite, whose eggs are safely inside. Some scelionids, whose larvae develop in grasshopper eggs, cling by their jaws to the underside of the abdomen of the full-grown female grasshopper. Their flattened, almost tick-like form help to keep them from being knocked off during this rough ride. They do not even loosen their hold at death in the collecting jar.

The remaining and best known Hymenoptera belong to the group Aculeata, the "sharp-tailed" insects with terminal stings (Fig. 69). The ovipositor, or sting, is no longer used to insert eggs in the host but to inflict stab wounds or to inject poisonous and pain-producing substances. It thus functions in killing or immobilizing prey or in defense.

Except for the gall-makers, who subtly influence their host plant by chemical means, the Hymenoptera that have been so far dealt with take nature about as they find it. The aculeates, however, are industrious insects that diligently modify nature. They construct elaborate shelters. Some grow crops and tend insect livestock. Such activities reach their height in the social Hymenoptera, where the combined efforts of diverse individuals structurally adapted for different tasks control their environment to a considerable extent.

There are about as many classifications of the aculeates into major groups as there are authorities on the subject, but one as good as any for a general discussion is a division into three groups: wasps, ants, and bees. The wasps are the predators, the ants are the wingless social forms, and the bees are the pollen-eaters. Some, at least, of the exceptions to these generalizations will be duly noted.

So far as can be determined, the true wasps arose from the parasitic Hymenoptera. The difference between the parasitism of the ancestral forms and the predation of the true wasps is not a great one, and a transition between the two modes of life can easily be imagined. The parasite puts its eggs on or in the host, then leaves both it and the young to their fate, although she may make things a little easier for her young by paralyzing the host with venom. The typical predatory wasp, on the other hand, after killing or paralyzing its prey, carries the victim away to a shelter that it has constructed, where the young can devour the prey in comparative safety. There are also other differences between "parasite" and "predator": the predatory wasp larva is not an internal feeder, as are many parasites, and it may be larger than its prey, devouring several individuals before becoming full-grown.

An informal classification of wasps will be used here that divides them into (1) primitive wasps, in which the predatory mode of life is often not well established; (2) the vespid wasps, of which the best known kinds are social; and (3) the solitary, or digger, wasps.

On structural grounds, some of the several families of wasps now taken up are among the most primitive of the aculeate Hymenoptera, and the habits of some tend to bridge the gap between predation and parasitism. These wasps typically hunt on the ground, and so highly adapted are some to this mode of life that the female (who as in all wasps does the hunting) has become wingless and is ant-like in form.

The insects that are hunted by these primitive wasps often live in burrows. It is therefore unnecessary for the wasp to construct another burrow to shelter the young and its food supply. For example, the wingless wasp *Methoca* preys on the larvae of tiger beetles, which live in short vertical tunnels. These beetle larvae seal the burrow entrance with their heads and thorax and capture with their powerful jaws insects that happen to walk by. The wasp runs about over the area inhabited by the tiger beetles until she is seized by one of them. Her armor withstands the trap jaws, and she stings the would-be predator in the neck, places an egg on it, and stuffs it back into its own burrow.

Large wasps of the family Scoliidae dig in the ground or follow tunnels after the grubs of scarab beetles. These grubs usually feed near the surface. After paralyzing the grub, the wasp, rather than leaving the victim in its own tunnel, digs some distance farther down and constructs a new cell for it. The practice of improving or enlarging the native dwelling of the prey, whether burrow or natural crevice, could lead gradually to the situation where the wasp did all of the construction. This would widen the range of prey to include free-living insects.

The most extensive group in this series of wasps is the family Psammocharidae, that of the spider wasps. Nearly all of these active, long-legged insects prey on spiders. When the prey is a spider that lives in a tunnel, it may be stored in its own dwelling after it is paralyzed and an egg placed on it. Most often the spider wasp digs a nest to receive the prey. In some species the wasp only temporarily paralyzes the spider, and its larva feeds on the living and active host in the manner of a parasite. One wasp no longer catches spiders itself but hunts out the nests of other species to replace the owner's egg with its own. Yet others oviposit in spiders that are being dragged home by a successful hunter; the parasite larva hatches first to destroy the egg of the original owner. The very large *Pepsis formosa* of the American Southwest attacks fist-sized ground spiders called tarantulas. These "tarantula hawks" overpower and paralyze one of these large spiders for each of the young.

Among the wingless primitive wasps the most familiar are the velvet ants, of the family Mutillidae. They usually are larger than ants and may be covered with long, brightly colored hairs. Their sting is unusually powerful; one species of the Southeast is known as the "cowkiller." Mutillids are especially common in open sandy places, where, in the cool of the morning or evening, the females may be seen running about in search of the nests of bees and wasps on whose larvae her young will feed. The females in some species are themselves carnivores and systematically hunt down wasps and bees, including the hive bee. The winged males feed on nectar.

The primitive wasps of the family Chrysididae, the cuckoo wasps, are jewel-like insects, with shining heavy armor of brilliant metallic blue or green. They can curl up tightly into a ball by bringing the hollowed-out underside of the abdomen up over the face. Most of these parasites invade the dwellings of other stinging Hymenoptera— usually solitary wasps, sometimes bees—where they leave a fatal egg. Usually the young cuckoo wasp delays its attack until the larva of the host has become full-grown, then devours it; or in some instances it destroys the victim at the beginning and itself devours the store of food.

All but a few of the most primitive members of the family Vespidae fold the wings lengthwise when not in use; this, together with technical aspects of wing venation and the structure of the thorax, defines the group structurally. Biologically, most are hunters of caterpillars; and it is in this family that the only social wasps are found.

Of interest to the student of insect societies is the transition between the solitary and the social mode of existence that occurs within the Vespidae. Most of the species are solitary, but some live in family groups, others are organized into rude societies of a few score inhabitants, and some live in societies of thousands of individuals, housed in a nest constructed by social labor.

The solitary, or nonsocial, species construct simple chambers which are individual shelters for the young. These cells may be dug in the earth, be partitioned off in hollow stems, or fashioned from mud that is plastered on stones or stems. After the cell is filled with the number of caterpillars that are needed to bring the young to maturity, an egg is placed in the cell, and the cell is sealed and opened only when the larva completes its development and the mature wasp digs its way out. After depositing the egg and sealing off the cell, the parent has nothing more to do with her offspring.

When, however, as in the African genus *Synagrus*, the egg is laid in the empty cell before provisioning starts, a situation is created in which a family unit appears. The larva hatches before the stock of food is laid in, so that the female has to feed her young from day to day. This parental care, or progressive feeding, establishes a family group.

The step from the family group to the true insect society is not a great one. All that is required is that some of the female offspring be inhibited from becoming sexually mature. It is known that in some social Hymenoptera this inhibition is accomplished by nutritional means, either by controlling the diet or by secretion of a hormone which the parent feeds them. The parent who deals with her daughters in this way achieves the status of queen. The appearance of the sterile caste marks the appearance of the insect society. As evolution of the different castes proceeds, each may add new adaptive structures and behavior patterns suited to its mode of life. Periodically, usually once a year, the inhibition is relaxed so that sexually mature females appear, to become the queens of next year's new colonies. As in all Hymenoptera, males are produced from unfertilized eggs. They always become sexually mature, never have any function other than reproducing, and do not become members of the sterile working caste.

Although not burdened with the complexities of life in a society, the solitary vespids face problems that require complex behavior for their solution. The ground-nesting species must learn to find the entrance to their nest after long hunting forays. The construction of the mud nests—often well designed, symmetrical structures—may be an elaborate operation. One solitary wasp (*Odynerus herrichi*) hunts an arboreal caterpillar that conceals itself under a web of silk. The wasp prepares for the capture by first cutting a hole in the web. She then prods the web with her sting until the caterpillar is flushed out. It emerges from the hole cut for it and falls to the ground. Now the wasps also plummets to the ground and searches for the fallen caterpillar. Should she fail to find it the first time, she flies back up to the web and makes another try by again allowing herself to fall.

The social wasps common in the United States belong to two genera, *Vespula* and *Polistes*. The species of *Vespula*—hornets and yellow jackets—build globular paper nests that hang in trees or in cavities underground. These wasps defend their colonies fiercely. People are most often stung by the ground-nesting species, for the nest entrance is not conspicuous and may be trampled on unknowingly.

In the tropics the nest of the social wasp may be a permanent affair, but in the North each nest is started anew in the spring and is abandoned in the fall. Nest construction is begun by a single overwintering female, the queen. She scrapes off mouthfuls of weathered wood with her mandibles, carries the material to the nesting site, and with the aid of an oral secretion converts it into pulp that is squeezed out into the gray sheets of paper used to build the nest. The brood cells are arranged in horizontal layers, face downward. In each of the first few cells the queen places a single egg, and she cares for the first brood of young. These young become sexually immature females who take care of the succeeding young and enlarge the nest. New layers of cells are added, one below the other, and the outside of the nest is insulated with several enveloping layers of paper. Sexually mature males and females appear in the fall. After mating, the females go into hibernation for the winter. Males, workers, and immature larvae die with the onset of cold weather.

Wasps of the genus *Polistes* (Fig. 70) are more often seen than are the hornets, because their small paper nests, consisting of a single exposed shelf of cells hanging from a short stalk, are common about buildings. The queen makes a dozen or so cells early in the spring, and in each places an egg and a few drops of thick honey. After the larvae hatch, she feeds them both honey and chopped caterpillar. The queen is not easily distinguishable from the workers, and some of the workers lay eggs (infertile ones that develop into males), so that caste differentiation is not strong. When the larva, hanging head down in its paper brood cell, is fed, it exudes a drop of secretion from the mouth, which is drunk by the nurse or queen wasp. This exchange of food, called trophallaxis, is evidently a mechanism that serves to hold the colony together. At maturity the grubs spin a cap of silk over the cell and pupate inside.

Polistes uses nectar to a limited extent in its economy, but a few other vespids, in the subfamily Masarinae, are like bees in that they rear their young entirely on a mixture of pollen and honey.

The sphecoid wasps—the solitary, or digger, wasps—are a diverse group of many families and some thousands of species that mostly dig their nests in the earth. Since dampness and mold are the worst enemies of such ground nesters, wasps abound in dry regions. Some make nests above ground of mud or nest in wood. They ceaselessly scour the earth for insects and spiders to feed their young.

Members of what is perhaps the most primitive family of sphecoid wasps, the Ampulicidae, gather cockroaches, which are themselves primitive insects. After delivering a sting that only quiets, not paralyzes, the cockroach, the wasp nips off a length of each antennae and drinks the blood that oozes out. Then, seizing the stump of an antenna with her jaws, the wasp leads the groggy cockroach to some natural crevice that serves as a nest for protecting the wasp larva which feeds on the cockroach.

FIG. 70.   *A wasp of the family Vespidae. It constructs hanging nests made of wood-pulp paper.*

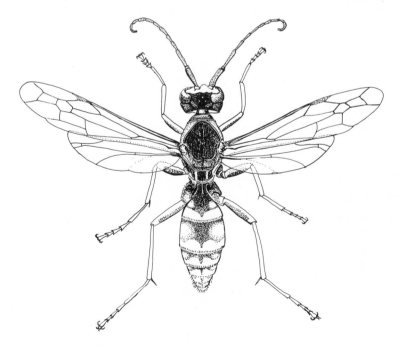

Some digger wasps restrict their prey to a single species of insect, or even to only one sex of the species. The nests of certain species of *Philanthus* are stocked only with brilliant green bees, and a European species preys on the domestic honeybee. Other wasps may fill their nests with a miscellany of insects of several orders.

The wasp may kill the prey with the sting or only paralyze it, either temporarily or permanently. If the prey is small, it may be unharmed, and, if the nest of the wasp is opened, the victims swarm out to escape.

The hunting wasps have been favorite subjects for students of insect behavior, who admire the strategems used to get the prey home—the cicada wasp that carries its bulky victim up a tree, flies as far as it can in a descending track, then climbs again to repeat the process; or the wasp that clips the wings off its prey for easier handling. Observers also describe the limited mental capacity of their subjects—the wasp's befuddlement when landmarks around her nest are altered, for example. But consider what the reactions of the experimenter himself would be should some incomprehensibly gigantic intruder stir up the landscape around his home.

Of the million or so species of animals known, only a few use tools: man; a finch of the Galapagos that uses a thorn to dig grubs out of bark; a sea otter that cracks shells with stones; and so on. One of them is a sphecoid wasp, *Ammophila*, that uses a pebble to flatten the earthen plug that seals her completed nest. She picks up the pebble in her jaws and bobs up and down with it at a furious rate, pounding away at the ground.

Where the other social Hymenoptera are winged, large-eyed creatures of the sun and flowers, the ants are earth-bound skulkers. They doubtless began their evolutionary history as roving bands of predatory wingless wasps, and some of the primitive hunting ants of Australia are little more than this. From such beginnings they have become the hymenopteran equivalent of termites, living in fixed and populous colonies and gathering plant and animal material—healthy, crippled, or dead. Ants must be regarded as the most successful and diversified of the social insects, since they abound in all regions, and their four thousand species are more than the social bees, social wasps, and termites combined. As in the termites, there is no spectrum of solitary forms grading into social ones: all living ants are social.

The individuals most often seen are wingless females, who belong to the nonreproductive worker caste. Only the reproductives—the male, who dies after mating, and the female, the founder and queen of the colony—are winged. These winged forms come out of the nest in swarms at certain times of the year, their emergence synchronized so that males and females of the same species from numerous colonies are flying at the same time.

After mating, the female sets about establishing a new colony. She bites off her wings, now merely impediments, and seeks out a shelter that will serve as a nucleus for a nest. Her wing muscles disintegrate and, together with the material stored in the fat body, are converted into a salivary secretion that is fed to the grubs hatching from her first batch of eggs.

The larvae develop into sterile, wingless females, or workers, that gather food, enlarge the nest, feed the young and the queen, and defend the colony. They and the following young may be all alike or may be structurally diverse, some with remarkably large heads and large jaws. When the difference between the small- and large-headed workers is pronounced, the latter are termed soldiers and may actually be important in the defense of the nest, although they also help with ordinary domestic duties.

Within the colony, ants continually feed each other. Even the larvae exude substances eagerly licked up by the workers. If a worker ant is fed a drop of deeply colored sugar water, in a matter of hours the pigment can be found in most of the ants in the nest. This socializing device, the exchange of nutrients (trophallaxis), is taken advantage of by a swarm of parasites and guests from many different groups of insects who, paying as the price of admission a small amount of tasty secretion, have at their disposal the larvae and the food supply of the ants.

Some tropical primitive ants are nomadic carnivores that move from camp to camp and kill everything in their path that they can overcome. It is said that even tethered domestic animals have been thus killed and torn to shreds. Most of the familiar ants of temperate regions are scavengers, bringing in dead insects, although killing some.

Many ants depend to a very great extent on the honeydew produced by aphids and related insects. The relationship between certain aphids and ants is so close that it is doubtful that the aphid could exist without the ant. Some species of ants store aphid eggs during the winter, then carry the helpless young out to pasture on plant roots in the spring. The North American *Cremastogaster lineata* builds paper sheds over colonies of aphids.

Ants get sugary secretions also from flowers and from the extra-floral nectaries of some plants, which are situated at the bases of leaves. Several groups of desert ants the world over store honey in workers that are fed until their abdomens swell like balloons. These honeypots hang from the roof of the nest by their legs and dispense drops of honey for the other ants of the colony on request.

The leaf-cutter ants (*Atta*) of the New World tropics base their economy on green leaves, which are inedible, but can be fed to sub-terranean gardens of fungi that are grazed on by the ants and given to their young. From the populous colonies of these large ants beaten pathways radiate out into the forest, and on them may be seen pro-cessions of workers carrying home like banners the pieces of green leaves. When the queens leave the nest on their mating flight, they carry with them in a pouch in the head a bit of the fungus garden for seed when a new colony is founded.

Members of the worker caste in the ant genus *Polyergus* are not really workers; they are unable to forage, construct nests, or feed their young. These needle-jawed Amazons are warriors and raid the nests of other ants, slaughtering the workers and bringing home alive the larvae and pupae of the victims. When the slave ants emerge from their pupal slumber, they take care of the young of the Amazon queen as if they were their own species.

Asiatic ants of the genus *Oecophylla* employ child labor in construc-ting their nests, which are made of leaves cemented together. One would think that the eyeless, legless maggots of the ant would be poor subjects for exploitation, but workers carry them in their jaws to the construction site, and, while other workers hold the leaf edges together, the larvae exude drops of liquid (used also by the larva to spin its cocoon) that harden into an effective cement.

Many ants have only rudimentary stings and, instead of stinging, eject vapors of formic acid. Others have fiery stings, as the inch-long bulldog ant of Australia and the fire ant of South America and the southeastern United States.

Nests of ants differ from those of other social Hymenoptera in that there are no individual brood cells for the young, which are reared in communal chambers. Typically, nests are irregular galleries dug in the ground, or sometimes in wood, but some ants build small paper nests in trees in the manner of wasps. Small cavities, such as hollow grass stems or fallen acorns, form compact nesting sites for minute species of ants.

Ants are good subjects for observations and experiments on group behavior. Their remarkable methods for achieving social integration, although complex, are yet simple enough to be analyzed with some completeness. It is interesting that A. Forel, a leading European student of ants, was a professional psychologist and that W. M. Wheeler, the famous American myrmecologist, had a deep interest in psychology.

Mark Twain was in error when he called bats Coleoptera, but it must be admitted that he had a good eye for ants. In *A Tramp Abroad*, he writes:

During many summers, now, I have watched him, when I ought to have been in better business, and I have not yet come across a living ant that seemed to have any more sense than a dead one. I refer to the ordinary ant, of course: I have had no experience of those wonderful Swiss and African ones which vote, keep drilled armies, hold slaves, and dispute about religion. . . . He goes out foraging, he makes a capture, and then what does he do? Go home? No, he goes anywhere but home. He doesn't know where home is. His home may be only three feet away—no matter, he can't find it. He makes his capture, as I have said; it is generally something which can be of no sort of use to himself or anybody else; it is usually seven times bigger than it ought to be; he hunts out the awkwardest place to take hold of it; he lifts it bodily up in the air by main force, and starts; not toward home, but in the opposite direction; not calmly and wisely, but with a frantic haste which is wasteful of his strength; he fetches up against a pebble, and instead of going around it, he climbs over it backwards dragging his booty after . . . comes to a weed; it never occurs to him to go around it;

no, he must climb it; and he does climb it, dragging his worthless property to the top . . . he lays his burden down . . . then . . . starts off in another direction to see if he can't find an old nail or something else that is heavy enough to afford entertainment and at the same time valueless enough to make an ant want to own it.

The bees are little more than furry wasps that have forsaken carnivorous ways to gather nectar and pollen. The mode of transition from wasp to bee has not been worked out to everyone's satisfaction. Some of the social wasps bring in honey, gathered from flowers and carried in their crop, and store it in the nest. Perhaps some of the solitary wasps, which are believed to be the ancestors of the bees, also behaved in this fashion (but if so, they have become extinct, since none are known to do this). A next, and also hypothetical step in the transition would be that the wasp eats pollen as well as nectar and regurgitates this mixture at the nest to be eaten by the young. What are believed to be the most primitive of bees belonging to the genus *Prosopis* (or *Hylaeus*) are nearly hairless insects that actually do carry pollen and nectar to their nest in this fashion. In a further stage of development, the bee would evolve body hairs that could trap and carry larger loads of pollen. Pollen is a rich food, comparable to meat in protein content, so that the transition from a carnivorous to a polleneating mode of life would probably not be difficult physiologically.

The external pollen-carrying devices of bees are varied. In the majority there are brushes of long hair on the hind legs or on the underside of the abdomen. These pollen-transporting structures differ in detail from one group of bees to another and are of great importance in the classification of bees. Also their minute structure is sometimes explicable in terms of the structure of the pollen grains they carry; thus, the hairs of the brush may be stiff and widely spaced when the pollen collected is of large, loose grains or may be fine and closely spaced for dust-like pollen. Even the structure of the individual hairs of the pollen brushes varies from group to group and may also be related to the structure of the pollen.

An advanced minority of the bees have hairless basins on the outer faces of the hind legs for carrying pollen. These are the social bees: bumblebees, honeybees, and related species. Such pollen-transporting

FIG. 71. *The pollen-carrying brush, or scopa, of the anthophorid bee is situated on the hind leg.*

baskets (corbiculae) are not specialized for carrying any particular kind of pollen; rather, they are of general rather than special usefulness, since the bee can pack almost any kind of pollen into them. The many species of parasitic bees have lost the pollen-carrying structures. They are therefore sometimes very much like wasps in appearance, and even the specialist has been misled into describing a wasp as a new species of parasitic bee. Usually a microscopic examination of these parasitic bees will disclose some branched hairs; these do not occur in any of the sphecoid wasps, the wasps superficially most like the bees.

Inside the thorax of the bee is a set of levers that aid in flexing the wings over the back after a landing. This structure, absent in wasps,

FIG. 72. *The megachilid bee carries pollen in a large brush on the underside of the abdomen.*

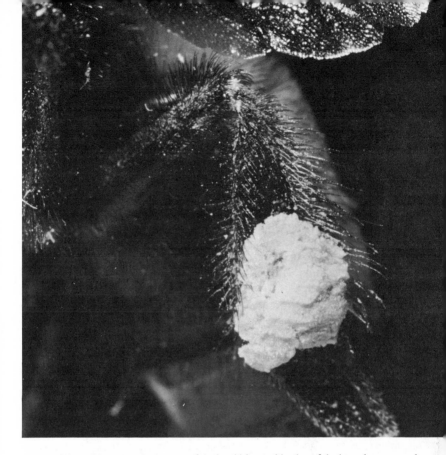

FIG. 73. *The pollen-carrying apparatus of the bumblebee is, like that of the honeybee, a smooth basin, margined with stiff hairs, on the outer face of the tibia of the hind leg. This pollen load is incomplete.*

is probably of adaptive value to the bee in its frenetic tour from flower to flower. From the sphecoid wasps the bees differ also in lacking a cleaning organ at the base of the hind tarsus. Both bees and wasps have a similar organ on the front legs, a device that neatly brushes debris off the antennae, but the bees have either discarded the corresponding structure on the hind legs or never had it.

In the closely intertwined evolutionary history of the bees and the flowering plants, those structures of the bees that have been most strongly affected are the pollen-carrying structures, already briefly described, and the tongue, fashioned from the labium and maxillae, which is used to drink nectar. Nectaries placed openly in shallow flowers are accessible to bees with short tongues, but those at the end of long corollas can be reached only by bees with specialized elongate tongues. Bees of several genera use the tongue to extract pollen from flowers of the genus *Cryptantha*, in which the anthers are hidden in a narrow, nearly closed corolla tube. These species all have the tongue fitted with recurved hooks that drag out the pollen grains. Closely related species of bees that do not visit these flowers have tongues of the normal, unarmed type.

All but the parasitic bees provide food for their helpless, grub-like young, either by leaving with the egg a lump of pollen mixed with nectar or, as in many of the social bees, by giving it food from day to day. This is, of course, done by the female, who alone is furnished with the pollen-carrying apparatus.

The young are reared in individual cells. Most bees are ground-nesters, digging a tunnel and furnishing it with several brood cells. Like the digger wasps, the bees flourish in dry regions. To protect the young from moisture and mold, some bees coat the cell walls with a secretion troweled on with the tongue. Bees of the genus *Megachile*, widespread in distribution and numbering thousands of species, are called leaf-cutters from their habit of carving out rounded pieces of green leaves. They carry these to their nest and ingeniously fashion them into linings and caps for the brood cells. Fossil leaves with the characteristic incisions of *Megachile* have been found in the Miocene shales of Florissant, Colorado. Some bees burrow in pithy stems or in wood and may line their nests with cottony masses shaved off hairy plant stems. Others build hard, exposed mud nests, and social bees make their nests of wax.

Where the nests of the digger bees are crowded closely into large colonies, the air hums with activity. Bees heavily laden with pollen and with their crops gorged with honey come winging in from the fields, swing in the air briefly above their nests, and abruptly drop out of sight. Drawn by this concentration of hosts are the many insects parasitic on the bees. Dingy gray tachinid flies lurk about the burrows; furry bombyliids, or bee flies, hover stationary in the air, then dart forward to place their eggs near the nests; and, deep in the burrows, minute larvae of beetles that rode in on the nesting bees scramble about in search of the young bees in order to destroy them and take possession of the pollen stores. Also among the parasites are other bees.

Besides the classic social bees—the bumblebees, the honeybee, and related species—there are two groups of mostly solitary bees that have evolved simple societies. Best known are some species of the ground-nesting genus *Halictus*, abundant, smallish bees with hard bodies and sharp stings. They are the sweat bees, that drink from perspiring skin and inflict a hot sting if one is careless in brushing them off. In the social species of *Halictus* the queen in early spring digs a nest with several brood cells and stores them herself. The young are females but differ from the parent in minute details of external structure and also are infertile (there are no males about, anyway). They have the pollen-gathering and cell-constructing instincts of their parent and care for the additional young she produces throughout the summer. In the fall the fertilized eggs develop, not into workers, but into sexually mature females who will become the queens of the following spring, and in the fall the queen presumably lays also some unfertilized eggs, for at this time numbers of males appear.

The "true" social bees, that comprise the family Apidae, are marked structurally by the pollen-baskets already mentioned, and biologically they are characterized not only by their sociality but also by their ability to secrete wax in such quantity that they can use it to fashion brood cells and excellent containers for pollen and liquid honey. This use of wax ranks as one of the great industrial achievements of insects.

The familiar bumblebees of the north, who work earnestly early and late and exist even in the Arctic, live in small ephemeral colonies that die each fall. The queens born at the end of the life of each colony mate and overwinter, then in early spring look for a nesting site—a warm and dry mouse nest or tuft of grass or a shelter under boards. In it she constructs a waxen honey jar, which she fills against the inclemencies of early spring days, when the weather keeps her indoors. In the nest, warmed by her body, her first young, which become smaller replicas of herself, quickly reach maturity. But they do not become sexually mature; their development stops short of mating and laying eggs. The young are now cared for by these workers. It is not until the end of the season that males and sexually mature females are again produced.

Occasionally the queen of one species of bumblebee will enter the nest of another species, kill the queen, and lay her own eggs there, so that the workers of the alien species rear her young. Evidently the parasitic bees of the genus *Psithyrus* originated in this way. They look very much like bumblebees but have become so adapted to the parasitic mode of life that they have lost the pollen-carrying baskets on the hind legs. The sting of the parasite is more powerful, and her armor tougher than that of the queen of the true bumblebee that she destroys when she victimizes a colony. With the rightful queen gone, the workers now have only the eggs of the *Psithyrus* to rear.

The flower of red clover has deep-seated nectaries that can be reached only by the long tongues of certain bumblebees, hence is pollinated only by them. When the red clover was introduced into New Zealand, it did not set seed because there were no bumblebees in that isolated place. Bumblebees were accordingly brought in, but too hastily, since both a proper long-tongued species and a short-tongued one were introduced, and, while the first did the job of pollinating as expected, the latter species, because it chewed holes in the flowers to get at the nectar, turned out to be harmful.

The abundant stingless bees, a group of more than two hundred species, are widespread in the tropics. There are two genera: *Melipona*, in the New World only, and with species honeybee-like in size and appearance; and *Trigona*, of all tropical areas, with species that are so small that some are called "mosquito bees." The South Americans call the stingless bees "Angelitos" because of their welcome deficiency, and, before the Europeans brought the honeybee to the continent, the stingless bees were kept for their honey in hives made of hollow logs.

The stingless bees defend their nests bravely, biting their enemies and smearing them with a sticky substance. Their nests, usually in hollow trees, are elaborately fortified with wax or clay and can be entered only by a long and narrow wax spout that is well manned by defenders.

A rather surprising difference between honeybees and the stingless bees is the fact that the latter do not feed the young from day to day but rather fill their cells with the requisite pollen and nectar and seal them off to complete their development alone. They are the only social Hymenoptera that do not feed the larvae progressively.

The society of the honeybee is a powerful one; the roar of an angry hive is most impressive. It is not easy to conquer by direct attack, and the quiet stratagems of the beekeeper best overcome it. The concentrated firepower of the stings of the tens of thousands of individuals in the hive is enormous.

There are in the Old World four species of *Apis*, but only one of these, *Apis mellifera*, is established in Europe and is the true honeybee. When the colonists brought the honeybee to North America, it invaded the woodlands ahead of the settlers. Thoreau wrote:

The honey bee hummed through the Massachusetts woods, and sipped the wild flowers round the Indian's wigwam, perchance unnoticed, when, with prophetic warning, it stung the red child's hand, forerunner of that industrious tribe that was to come and pluck the wild flower of his race up by the root.

In the hive are three quite different types of bees. By far the most numerous are the workers, the only type seen by all but the close observer. The workers are sterile females and are equipped with pollen baskets, wax glands, and large salivary glands that manufacture the equivalent of milk for the very young bees. In the hive there is usually only a single reproductively active female, the queen. She is larger than the workers, with a long abdomen that looks too large for the wings, and she can in fact fly only with difficulty when in good egg-laying condition. She lacks the pollen baskets and specialized glands of the workers. At certain times of the year there are numerous males, the drones, that have very large eyes and whose only activity is to eat and to participate in the mass flight of the males after the virgin queens. Only one out of many mates, and he dies in the process; the remainder are eventually ejected from the hive by the workers.

The queen has one visible function only. She lays eggs, dipping her abdomen into cell after cell, at the rate of hundreds a day. Thus, most of the ceaseless activity of the hive is carried out by the workers. Some are in the field gathering food from flowers, others are constructing cells of wax, others are feeding the young. These activities are usually going on simultaneously, as if under the direction of a reasonably competent organizer.

Apparently the factor that has been taken hold of by natural selection to attain this kind of organization is the age of the worker; that is, the young worker will have a behavior pattern that accomplishes one set of jobs, the older worker another pattern, and so on. There are, then, faint indications of the kind of polymorphism that is found in metamorphosis, but it affects mainly behavior and is confined to adult life.

When the worker first emerges from the brood cell, her duties are intimately connected with these cells. For the first three days of her life she cleans out used cells and varnishes them with her tongue. If this is not done, the queen will not lay in them. Then, until about the end of the first week, she feeds the older grubs that are to become workers; their food is a mixture of pollen and honey, which she obtains from storage cells.

During the second week of her life, the salivary glands become greatly developed, and she secretes and feeds to the grubs a special food called royal jelly. Only the very young worker grubs get this, while the grubs that are to become queens are fed it throughout larval life.

At the beginning of the third week, the wax glands of the worker develop, and from the plates of wax that protrude from the underside of the abdomen she fashions new comb, which consists of a double layer of beautifully constructed hexagonal cells, placed back to back. The cells are all alike, except for some large ones built for the larvae of drones, and are used both for storing food—honey and pollen—and for rearing young. The worker of this age also takes the loads of pollen and nectar from the field bees as they arrive and transfers the material to the storage cells. Also she makes practice flights around the hive, memorizing the landmarks in preparation for the last phase of her existence, that of the field bee.

At the beginning of her fourth week and until her death some twenty days later, the worker ranges far afield in search of water, nectar, pollen, and propolis, a plant resin used to patch cracks in the hive.

This pattern of behavior is not absolutely rigid, and, when the hive becomes deficient in certain age groups, older bees may take over the work normally done by younger bees.

If a few broken honeycombs are placed in the open, they at first lie unattended, then a wandering bee finds them, drinks, and disappears. In a few minutes several, then hundreds of excited bees crowd to the scene. Done on a more delicate scale, by putting out a small dish of scented sugar water, one sees again a long period of neglect, the single visitor, then the sudden influx of bees. Obviously, the discoverer has returned to the hive and led an expedition back to the sugar water.

It might be worth-while to check the theory, however. Dab a little paint on the back of the first bee, and observe how she brings back her companions. But the painted one may not reappear with the small flock of bees that come. She did not lead them, she told them.

This relates, of course, to the famous observations of K. von Frisch that made known to the world the language of bees. The painted bee arrives at the hive, importantly full of sugar water, and the other bees crowd around her. From her scent they note what she has found. She begins an excited dance, circling to the right, then to the left, at each turn wagging her abdomen. From this dance, the watching bees discover where she found the syrup.

By placing the bait at different distances and directions from the hive and observing the changes in the behavior of the dancing bee, the dance was decoded. When the dish of sugar water is far away, the bee dances more slowly than if it had been placed nearer, with fewer turns in the dance for each time unit, thus giving the distance. The direction taken by the bee during the wagging portion of the dance indicates the direction of the food. Since the dance is performed on a vertical comb, the bee can not point directly. Rather, she speaks in terms of gravity and relative position of the sun. If the bee wags its abdomen while walking upward in its dance, this means that the sugar is in the direction of the sun; if while walking downward, it is directly away from the sun; to the right or left, to the right or left of the sun. With these instructions, the bees that observed the dance can fly directly to the general vicinity of the dish of sugar water, where by cruising about they can locate it exactly by scent and sight. If the bait is placed near the hive, within a hundred yards or less, the direction-indicating part of the dance is omitted, and the searching bees merely fly at random about the hive until they find it.

When a bee gets into the wrong hive, it is pounced on and ejected. The stranger is recognized by its odor. The distinctive odor of each hive results from the mixture of pollen and nectar peculiar to it; the hives of bees in an area are not uniform in their flower-visiting habits at any given time. The uniqueness of each hive is apparently in the long run based on the activities of a relatively few bees—the innovators, pioneers, or maladjusted individuals—who are prowling about while the mass of the field bees are exploiting proven resources. Should the scout bee happen to find a rich source of food, its excitement or persuasiveness might convert the majority to its discovery. The foraging activity of a hive is thus dependent to a degree upon the accidents of discovery on the part of its scout bees, so that two hives standing side by side may at the same time be exploiting different kinds of flowers lying in different directions.

The society of the honeybee, unlike those of other social bees and the social wasps of temperate regions, is a permanent one, capable of indefinite existence. Enough honey—perhaps fifty pounds or more—is sealed in the wax combs to keep the queen and workers alive and warm through the winter. Very early in the year, before the first flowers are out, the bees break open their pollen stores to feed new young of the year in order to have a strong and fresh working force to meet the flowering season. Although with potential immortality, the colony is in fact liable to destruction through disease, attack by predators, unfavorable weather, and so on, and the species would not survive if it were not for the fact that new colonies are founded.

The queen is so highly specialized for life as an egg-laying machine that she is quite unable to build the comb and gather the food necessary to start a colony. The honeybee colony therefore reproduces by dividing, a large number of workers accompanying the queen on a swarming flight to a new location.

When a hive is preparing to divide, or swarm, the workers construct several large brood cells and, by feeding the grubs on the royal diet, rear them to become queens. As the new queens, one of whom will inherit the hive, become mature the old one prepares to leave. Scout bees have gone out to find a suitable shelter for a new hive. The queen flies heavily for a short distance, then comes to rest on a low branch. The workers who are to leave gather about her in a heavy, hanging cluster. On the following day the assembled migrants take wing and fly to their new home, where the workers at once begin to transform the honey that filled their crops into the snow-white wax of the new combs.

In the old hive, left with a number of older workers, a few queens in cells, and larvae in all stages of growth, there is staged the battle of the queens. With the royal, high piping note, the first queen out of her cell tears open the cells of her sisters, and they fight with their stings. Death from the sting is sure, so that there are no wounded and only a single survivor. The surviving queen, in her earliest youth a powerful flyer, now takes wing on her mating flight. A train of drones follow her into the sky. The one to reach and capture her high in the air is killed in the act of mating, for it tears the viscera from his body. Once impregnated, the queen can lay fertile eggs for the rest of her life, as long as three or four years.

Until about a hundred years ago, it was impossible to manage bees on a large scale in anything like a scientific manner. They were kept in simple hives—shelters of straw or wicker—that were watched so that swarms could be captured before they got away. What went on inside the hive could be neither seen nor controlled. At the end of the season, the hives that were to be harvested for wax and honey were destroyed by suffocating the bees with burning sulphur. The rest were "seed" for the next year.

Beekeeping had hardly progressed beyond this primitive stage until an American, L. L. Langstroth, in the mid-nineteenth century solved the problem that had baffled students of bees for centuries—how to construct a hive with combs that could be removed and replaced at will. Even though one made a hive in movable sections, the bees glued the pieces together so firmly with propolis and wax that the hive had to be torn apart forcibly and was little better than the ancient straw hives, or skeps.

Langstroth observed that any space in the hive that measured between 3/16 and 3/8 of an inch was left open, while all those larger or smaller were filled in by the bees. This distance corresponds to the width of the natural passageways in the hive, particularly the distance between the faces of the combs. Langstroth made his movable combs so that they hung free in the hive, separated from the wall except at narrow points of contact by the proper distance. With such an arrangement it becomes possible, with the aid of a light prying instrument, to take combs out of the hive or put in new ones with minimum disturbance to the bees. In this way the size of the brood, the amount of stored food, and the condition of the queen can be observed readily, so that the experienced beekeeper can make the alterations necessary to keep the hive flourishing and producing the maximum amount of honey. The combs that are filled with honey can be harvested without destroying the colony.

Conclusion

Could we imagine as a single performance the story of animal life as it has been enacted on the continents of the earth, we could see it as a series of pyrotechnic displays: the slowly swelling burst of the reptiles showering out as the host of great and small dinosaurs, the winged pterodactyls, the water-dwellers, and falling as dying sparks, with some still glowing as the reptiles of today; the brilliant and briefer display of the mammals, its embers only now dying out in the savannas of Africa; and in the last instant the sharp explosive flash of mankind. The great burst of insect life would reach its maximum just after the fading of the reptiles and would seem to spread out as an enormous shower of tiny particles that lie still living, as diverse and numerous as so many glowing snowflakes, over the face of the land.

The odd thing about the evolutionary history of insects is that they seem to be essentially the same today in structure, diversity, and abundance as they were at the beginning of the Cenozoic era. During the Cenozoic, while the mammals were evolving into a succession of remarkably diverse groups, which followed one another in extinction, the insects remained static.

On the shores of the Baltic Sea the waves wash out from the rocks pieces of resin from a great ancient forest that grew there about 50 million years ago (early Cenozoic). These bits of fossil resin, or amber, being only slightly denser than sea water, are readily washed ashore, and for centuries the clear, shining brown substance has been gathered as gem material. In the amber are insects so well preserved that they look as if they had been put in pine resin only yesterday. In general appearance, and sometimes in exact and specific detail, the insects are like those now living—there are ordinary looking ants, ordinary looking flies, and no spectacular extinct groups.

The most important events in the insect world since the time of the Baltic forest have been concerned, as nearly as we can tell, with changes in geographic distribution as the climate became harsher and more abruptly variable toward the end of Cenozoic times, producing near disaster in the great land masses of the Northern Hemisphere during the Ice Age. Thus, in the Baltic amber and in such rich fossil deposits as the Florissant shales in the mountains of Colorado, we find insect types that, although still living, exist today only in milder lands.

The world's greatest collection of amber insects, which was in the museum of Königsberg, was burned during World War II, but in existing collections the vast majority of fossil insect specimens are also from the Cenozoic, when little of evolutionary importance was going on in this group of animals. The crucial periods were the Cretaceous, when the modern plant world, and with it the modern insect world, were taking form, and the Devonian and early Carboniferous, when the insects were emerging as the only flying invertebrate animals. Fossil insects from rocks of these periods are rare or not yet known. It requires an extremely unlikely combination of events to preserve insects as fossils, but deposits of these ages may someday be found, and much more can then be said about the evolutionary history of the class Insecta.

An attempt has been made in this book to give the reader a glimpse of the richness of a sector of the living world. It often is said that we are now in the age of science, that this is an age in which science dominates all, and it is pointed out that most of the scientists who ever existed are alive today. But it may turn out that this is only a crude and trifling beginning of an age of science. The reason for thinking this is that current science is largely physical science, and the purely physical universe and the science that manages it, is in a way extremely simple as compared with the biological universe and the biological science that can be developed in relation to it. An investment of human effort in biology commensurate with that of the physical sciences and with the inherent complexity of biological materials would probably require the cooperation of most of the human population. There is room not only for abstract mathematical thinking but also for the warm intuitiveness of the gardener, the hunter, the observer of living animals. And above all, since man is part of the biological universe, there would be need for a nearly incomprehensible degree of humaneness, in the face of the development of techniques that can change man himself.

When one sees the wasteland produced by the extension of military and business ethics into the universities and research laboratories, one cannot contemplate with any pleasure the emergence of biology as a major activity if it is to be organized along similar lines. To have the spokesmen of haste, destruction, and profits take over the area that ties man to the living world that gave him birth would be an affront of a very special kind and significance. But so long as the life-wish rules the human heart, the poet Coleridge will remain the spokesman of the relation between man and nature:

> Beyond the shadow of the ship,
> I watch'd the water-snakes:
> They moved in tracks of shining white,
> And when they rear'd, the elfish light
> Fell off in hoary flakes.

> Within the shadow of the ship
> I watched their rich attire:
> Blue, glossy green, and velvet black,
> They coil'd and swam; and every track
> Was a flash of golden fire.

> O happy living things! no tongue
> Their beauty might declare:
> A spring of love gush'd from my heart,
> And I bless'd them unaware:

Sure my kind saint took pity on me,
And I bless'd them unaware.

The selfsame moment I could pray:
And from my neck so free
The Albatross fell off, and sank
Like lead into the sea.

Bibliographical appendix

This bibliographical appendix includes works that will be of interest to the general reader but goes considerably beyond this aim, citing references of use to the more advanced student. While it is by no means a complete bibliography of the major works in English on entomology, many of the works cited contain extensive detailed bibliographies, and some of the purely bibliographic reference works are themselves listed.

Among the best of the general, nontechnical books on insects is A. D. Imms, *Insect Natural History* (London, 1947), which is concerned mainly with the nonphysiological aspects of insect biology. A more detailed but nevertheless readable account of the insect world is that of David Sharp in Volumes V and VI of the *Cambridge Natural History* (London, 1895–1901). A rather out-of-the-way source for an interesting account of a wide range of insects is T. D. A. Cockerell, *Zoology of Colorado*, pp. 143–230 (Boulder, Colo., 1927). The writings of Fabre on the behavior of insects, based on observations made in southern France, are both literary and scientific classics; an introduction to them may be found in Edwin Way Teale's *The Insect World of J. Henri Fabre* (New York, 1949). *The Insect Guide*, by Ralph B. Swain (New York, 1948), is a popular account, emphasizing the recognition of common insect families of North America, and has excellent colored illustrations drawn by Su Zann Swain.

At the reference and textbook level, the recent revision of Imms's *Textbook of Entomology*, by O. W. Richards and R. G. Davies (London, 1957), and E. O. Essig, *College Entomology* (New York, 1942), are standard treatments of world-wide scope. Charles T. Brues, A. L. Melander, and Frank M. Carpenter, *Classification of Insects* (Harvard Museum of Comparative Zoology Bulletin; Cambridge, Mass., 1954), is a technical key for the identification of the families of insects and other terrestrial arthropods of the world; it contains also an extensive bibliography on insect classification and has an original and important section on the classification of fossil insects. Emphasizing the American fauna are such standard texts as Donald J. Borror and Dwight M. Delong, *An Introduction to the Study of Insects* (New York, 1954), and John H. Comstock, *An Introduction to Entomology* (9th ed.; Ithaca, N. Y., 1940). Differing from these four in placing less emphasis on classification and more on general insect biology, including physiological aspects, is Herbert H. Ross, *A Textbook of Entomology* (2nd ed.; New York, 1956).

A classic account of insect physiology is that of V. B. Wigglesworth, *The Principles of Insect Physiology* (5th ed.; London, 1951). This modern and expanding field is covered in part by a volume edited by Kenneth D. Roeder, *Insect Physiology* (New York, 1953). A relatively nontechnical compendium of insect food habits and interesting entomological facts in general is to be found in Charles T. Brues, *Insect Dietary* (Cambridge, Mass., 1946). The United States Department of Agriculture Yearbook for 1952, *Insects* (Washington, D. C., U. S. Govt. Printing Office), has a wide range of articles on insects. Varied aspects of development and structure, from the microscopic level on up, are dealt with very satisfactorily in a book restricted to a single insect, *Biology of Drosophila* (New York, 1950), edited by M. Demerec.

The world literature on insects as it appears in books and technical journals is brought together and cross-indexed in the annual volumes of the *Zoological Record* (London). The *Zoological Record* deals primarily with the literature that is not concerned with economic aspects of the field; these are covered in *Review of Applied Entomology: Series A, Agricultural* and *Series B, Medical* (London).

An extensive technical glossary of terms used in entomology is provided by J. R. de la Torre Bueno, *A Glossary of Entomology* (New York, 1937). A list of common names of economically important insects of North America is that of Jean L. Laffoon, "Common Names of Insects Approved by the Entomological Society of America," *Bulletin of the Entomological Society of America*, VI (1960), 175–211.

A short history of the study of insects is given by E. O. Essig, "A Sketch History of Entomology," *Osiris*, II (1936), 80–123. In a centennial volume celebrating the founding of the California Academy of Sciences, *A Century of Progress in the Natural Sciences, 1853–1953* (San Francisco, 1955), is a valuable series of essays on the development of knowledge about the classification of many of the insect orders. A "Bibliography of Biographies of Entomologists," prepared by Mathilde M. Carpenter, will be found in the *American Midland Naturalist*, XXXIII (1945), 1–116.

1. The place of insects in nature

One of the classics in the technical literature of evolutionary biology is the paper by Robert E. Snodgrass entitled *Evolution of the Annelida, Onychophora and Arthropoda* (Smithsonian Miscellaneous Collections, Vol. 97, No. 6; Washington, D. C., 1938). In it a large complex of data on development and comparative morphology is given clean-cut organization as it is used to work out a scheme of the evolutionary relationships of these animal groups. Careful study of this work will give the student a sound background on the structure of the arthropods. In a more recent work, *A Textbook of Arthropod Anatomy* (Ithaca, N. Y., 1952), the same author has presented the subject in a more routinely descriptive fashion.

Identification of the North American representatives of arthropod groups other than insects can be facilitated by the use of Henry S. Pratt, *A Manual of the Common Invertebrate Animals, Exclusive of Insects* (Rev. ed., Philadelphia, 1935). The most comprehensive yet detailed account of a spider fauna in the United States is that of B. J. Kaston, *Spiders of Connecticut* (Connecticut State Geological and Natural History Survey Bulletin No. 70; Hartford, 1948); B. J. and Elizabeth Kaston have written an identification manual for the commoner North American species, *How to Know the Spiders* (Dubuque, Iowa, 1953).

A general natural history of the non-insectan terrestrial arthropods is furnished by J. L. Cloudsley-Thompson, *Spiders, Scorpions, Centipedes and Mites; The Ecology and Natural History of Woodlice, "Myriapods" and Arachnids* (New York, 1958). A good modern treatment of the biology and classification of spiders is that of Willis Gertsch, *American Spiders* (New York, 1949). An account at a more popular level is that of John Crompton, *The Life of the Spider* (New York, 1951).

2. The diversity of insects

A comprehensive text which gives an insight into the problems and methods of classifying and naming animals is that of Ernest Mayr, E. G. Linsley,

and R. L. Usinger, *Principles of Systematic Zoology* (New York, 1953).
Some aspects of museum work in insect systematics are dealt with by John
Smart, "Entomological Systematics Viewed as a Practical Problem," in
Julian Huxley, *New Systematics* (New York, 1940), pp. 475–592.

3. *The outside*

A standard treatise on insect structure is Robert E. Snodgrass, *Principles
of Insect Morphology* (New York, 1935), which describes internal as well as
external structure. A similar but less detailed treatment of the subject, one
adapted for classroom use, is E. Melville DuPorte, *Manual of Insect
Morphology* (New York, 1959).

The structure, function, and developmental origin of the exoskeleton,
or cuticle, are dealt with by A. G. Richards, *The Integument of Arthropods*
(St. Paul, Minn., 1952), and the subject is brought up to date by V. B.
Wigglesworth, "The Physiology of Insect Cuticle," *Annual Review of
Entomology*, II (1957), 37–54.

The structure of the male genitalia of insects, which are important in
insect classification, is discussed by Robert E. Snodgrass, *A Revised Interpretation of the External Reproductive Organs of Male Insects* (Smithsonian Miscellaneous Collections, Vol. 135, No. 6; Washington, D. C., 1957); this
paper has a bibliography that will serve as an introduction to the literature
on that rather controversial subject.

4. *Flight*

In addition to the chapters on flight in standard entomology and insect
physiology texts already mentioned, there is an up-to-date treatment of
the subject, with an extensive bibliography, by J. W. S. Pringle, *Insect
Flight* (Cambridge, Eng., 1957).

5. *The inside*

The structure of the internal organs is covered by references on insect
morphology given in the bibliography for Chapter 3, and the functions
of these organs in the general introductory references on insect physiology.
The following more specialized papers also will be useful: D. F. Waterhouse,
"Digestion in Insects," *Annual Review of Entomology*, II (1957), 1–18; and
G. R. Wyatt, "The Biochemistry of Insect Hemolymph," *Annual Review
of Entomology*, VI (1961), 75–102.

6. *Reproduction and development*

Material included in this chapter is dealt with in the standard textbooks of entomology and insect physiology. A more specialized treatment is that of V. B. Wigglesworth, *The Physiology of Insect Metamorphosis* (Cambridge, Eng., 1954). The use of chromosomal structure and behavior in the work of taxonomists is dealt with by M. J. D. White, "Cytogenetics and Systematic Entomology," *Annual Review of Entomology*, II (1957), 71–90. Material on the evolutionary status and biological significance of the larval stage is given in F. I. van Emden, "The Taxonomic Significance of the Characters of Immature Insects," *Annual Review of Entomology*, II (1957), 91–106.

7. *Sense organs and behavior*

Again, the subject matter is dealt with in appropriate sections of the standard texts that have already been cited; here it should be noted in particular that insect behavior is dealt with in a series of articles by T. C. Schnierla in *Insect Physiology*, edited by Kenneth D. Roeder (New York, 1953). The works of J. Henri Fabre, previously mentioned, are rich in observations on insect behavior.

The rather complex but easily observed behavior patterns that occur in those wasps that provision their nests with food for the young are favorite subjects for students of animal behavior. Observations made on wasps in a dune region of Holland are used to illustrate modern approaches to problems of the behavior in Nikolaas Tinbergen, *Curious Naturalists* (New York, 1960). An older descriptive account based on American species is that of George W. and Elizabeth G. Peckham, *On the Instincts and Habits of the Solitary Wasps* (Madison, Wis., 1898). An example of the use of inherited behavior patterns in classification is to be found in H. E. Evans, "Comparative Ethology and the Systematics of Spider Wasps," *Systematic Zoology*, II (1953), 155–72.

8. *Climate and season*

A general account is that of B. P. Uvarov, "Insects and Climate," Entomological Society of London, *Transactions*, LXXIX (1931), 1–247. Diapause has been one of the most intensively investigated aspects of insect physiology; in addition to A. D. Lees, *The Physiology of Diapause in Arthropods* (Cambridge, Eng., 1955), such articles as those of H. G. Andrewartha,

"Diapause in Relation to the Ecology of Insects," *Biological Reviews*, XXVII (1952), 50–107, and of H. E. Hinton, "The Initiation, Maintenance and Rupture of Diapause: a New Theory," *Entomologist*, XCVI (1953), 279–91, may be cited, references with extensive bibliographies that will open up the subject to the interested reader. One of the aspects of diapause is discussed by R. W. Walt, "Principles of Insect Cold-Hardiness," *Annual Review of Entomology*, VI (1961), 55–74.

C. B. Williams, *The Migration of Insects* (London, 1958), is a comprehensive treatment of the mass movements of insects that often are made as an adaptation to seasonal change.

Discussion of the relative roles of climate on the one hand and predators, parasites, and disease on the other in determining or controlling the size of insect populations will be found in such works as H. G. Andrewartha and L. C. Birch, *The Distribution and Abundance of Animals* (Chicago, 1954), W. R. Thompson, "The Fundamental Theory of Natural and Biological Control," *Annual Review of Entomology*, I (1956), 379–402, and A. J. Nicholson, "Dynamics of Insect Populations," *Annual Review of Entomology*, III (1958), 107–36.

9. *Insects and plants*

A general reference of importance here is the already cited *Insect dietary*, by Charles T. Brues (Cambridge, Mass., 1946). G. S. Fraenkel has been active in putting forward the hypothesis that the plant alkaloids are essentially chemical defenses against insect attack; see "The Raison d'Etre of Secondary Plant Substances," *Science*, CXXIX (1959), 1466–70.

An important aspect of the relationship between insects and agricultural plants is thoroughly dealt with by R. H. Painter, *Insect Resistance in Crop Plants* (New York, 1951), and an interesting specific example of the role of alkaloids in plant resistance will be found in H. Teas, "Physiological Genetics," *Annual Review of Plant Physiology*, VIII (1957), 393–412.

A beneficial aspect of plant and insect relationships is dealt with by C. B. Huffaker, "Biological Control of Weeds with Insects," *Annual Review of Entomology*, IV (1959), 251–76.

10. *Insects and other animals*

Hugh B. Cott's *Protective Coloration in Animals* (New York, 1940) was one of the pivotal books in the renaissance of the Darwinian approach in biology;

it contains much on the adaptive significance of coloration in insects, a significance which lies mostly in the relationship between insects and animals which prey on them. The whole problem of the relationship between genetics (a science at first antagonistic to Darwinian theory) and the theory of evolution by natural selection is treated in a broad way by the fundamentally important work of T. Dobzhansky, *Genetics and the Origin of Species* (New York, 1937); as in other historically important books, the first edition is perhaps most interesting, best exhibiting the revolutionary aspects of the new approach. The importance of insects in providing materials for evolutionary studies is well demonstrated by the contents of this book.

A standard textbook on the insects of medical significance to man and domestic animals is that of William B. Herms, *Medical Entomology* (New York, 1950).

11. *Insects versus insects*

The intensive warfare waged by insect on insect is of extreme interest to man because thereby is the number of his insect enemies diminished. The literature on biological control—so-called to distinguish it from chemical control, in which insecticides are used to kill the pests—is therefore extensive. It will be sufficient here to mention a standard text on the subject, H. L. Sweetman, *Biological Control of Insect Pests* (Ithaca, N. Y., 1936), and a more recent review with the same title by C. P. Clausen, *Annual Review of Entomology*, III (1958), 291–310. An important general work on the biology of the insects that parasitize or prey on other insects is Clausen's *Entomophagous Insects* (New York, 1940).

12. *Insect life in the water*

Keys and illustrations for the identification of insects that live in the fresh waters of North America have been written by numerous specialists in the textbook assembled by H. B. Ward, G. C. Whipple, and W. T. Edmonson, *Guide to Freshwater Invertebrates* (New York, 1959). An important group is dealt with by H. B. Hungerford, "The Biology and Ecology of Aquatic and Semiaquatic Hemiptera," *University of Kansas Science Bulletin*, XI (1919). Various aspects of respiration of aquatic insects are taken up in the textbooks of insect physiology already cited, while W. H. Thorpe, "Plastron Respiration in Insects," *Biological Reviews*, XXV (1950), 344–90, discusses

in detail one interesting method of respiration. Older, but still useful, is
L. C. Miall, *The Natural History of Aquatic Insects* (London, 1895).

13. *Insects without wings*

Rhyniellia praecursor, the most ancient six-legged arthropod, is discussed
in detail by D. J. Scourfield, "The Oldest Known Fossil Insect," *Nature*,
CXLV (1940), 799–801. An extensive work on the living Collembola, the
most abundant of the wingless insects, is Elliott A. Maynard, *A Monograph
of the Collembola or Springtail Insects of New York State* (Ithaca, N. Y., 1951).

The general biology of a thysanuran is given in H. L. Sweetman, "Physical
Ecology of the Firebrat, *Thermobia domestica* (Packard)," *Ecological Monographs*, VIII (1938), 286–311.

Charles L. Remington discusses the relationships between the orders of
wingless six-legged arthropods as well as their relationship to the winged
ones in "The Apterygota," in *A Century of Progress in the Natural Sciences,
1853–1953* (San Francisco, 1955), pp. 495–505.

14. *Primitive winged insects*

Two papers which give a general introduction to the fossil insects—and
the fossil record is essentially restricted to the winged insects—are those of
F. M. Carpenter: "Early Insect Life," *Psyche*, LIV (1947), 65–85, and "The
Evolution of Insects," *American Scientist*, XLI (1953), 256–70. An interesting
paper discussing the origin of flight is that of William T. Forbes, "The
Origin of Wings and Venational Types in Insects," *American Midland
Naturalist*, XXIX (1943), 381–405. The two primary references for the
Paleoptera-Neoptera theory are A. B. Martynov, "Über zwei Grundtypen
der Flügel bei den Insekten und ihre Evolution," *Zeitschrift der Morpholo-
gische und Ökologische Tiere*, IV, No. 3 (1925), 465–501, and G. C. Crampton,
"The Phylogeny and Classification of Insects," *Pomona Journal of Entomology
and Zoology*, XVI (1924), 33–47. A recent conception of the phylogenetic
relationships of the orders of insects is Herbert H. Ross, "The Evolution
of the Insect Orders," *Entomological News*, LXVI (1955), 197–208.

There is no modern general work on North American Orthoptera, but
W. S. Blatchley, *The Orthoptera of Northeastern America* (Indianapolis, 1922),
M. Hebard, "The Dermaptera and Orthoptera of Illinois," *Illinois Natural
History Survey Bulletin*, XX (1934), 125–279, and Gordon Alexander, "The
Orthoptera of Colorado," *University of Colorado Studies*, Series D, I (1941),
129–64, serve as regional studies. B. P. Uvarov, *Locusts and Grasshoppers*

(London, 1928), deals with some of the major plague locusts. C. V. Riley has a historically interesting account of the locust plagues in the early American West in the *First Annual Report of the United States Entomological Commission for the Year 1877 Relating to the Rocky Mountain Locust* (United States Geological Survey; Washington, D. C., 1878). A recent account of the gregarious locusts is D. L. Gunn, "The Biological Background of Locust Control," *Annual Review of Entomology*, V (1960), 279–300.

The biology of termites is discussed in such accounts of the social insects as W. M. Wheeler, *The Social Insects, Their Origin and Evolution* (New York, 1928), and O. W. Richards, *The Social Insects* (London, 1953).

15. *A specialized sideline*

As with most of the major orders of insects, there is no adequate treatment of the Hemiptera of the United States, but as a substitute one can, for a general idea of the diversity of the group, refer to T. R. E. Southwood and D. Leston, *Land and Water Bugs of the British Isles* (London, 1959). An identification manual for many of our species is W. S. Blatchley, *Heteroptera or True Bugs of Eastern North America* (Indianapolis, 1926); another useful manual which covers both Hemiptera and Homoptera is that of W. E. Britton, *The Hemiptera or Sucking Insects of Connecticut* (Connecticut State Geological and Natural History Survey, Bulletin No. 34; Hartford, 1923). An account of recent developments in our knowledge of the biology of an interesting and important group of homopterans, the aphids or plant lice, is given by J. S. Kennedy and H. L. G. Stroyan, "Biology of Aphids," *Annual Review of Entomology*, IV (1959), 139–60. An interesting paper on the periodical cicadas is that of R. D. Alexander and T. E. Moore, *The Evolutionary Relationships of 17-Year and 13-Year Cicadas, and Three New Species (Homoptera, Cicadidae, Magicicada)* (Miscellaneous Publications of the Museum of Zoology, University of Michigan, No. 121; Ann Arbor, 1962). Monographic treatments of the minor parasitic orders presumed to be related to the Hemiptera are G. F. Ferris, *The Sucking Lice* (Pacific Coast Entomological Society Memoir 1; San Francisco, 1951); and L. Harrison, "The Genera and Species of Mallophaga," *Parasitology*, IX (1916), 1–156.

16. *Ancient aquatics*

The North American stonefly fauna is described in two monographs on the mature and immature stages: J. G. Needham and P. W. Claasen, *A Monograph of the Plecoptera or Stone-Flies of America North of Mexico* (Thomas

Say Foundation, Publication 2; Lafayette, Ind., 1925), and Claasen, *Plecoptera Nymphs of America (North of Mexico)* (Thomas Say Foundation, Publication 3; Springfield, Ill., 1931).

An important regional study of the mayflies is that of B. D. Burks, "The Mayflies, or Ephemeroptera, of Illinois," *Illinois Natural History Survey Bulletin*, XXVI (1953), 1–216. The biology of these insects is dealt with by J. G. Needham, J. R. Traver, and Yui-Chi Hsu, *The Biology of Mayflies* (Ithaca, N. Y., 1935). An account of one interesting aspect of their biology is to be found in George F. Edmunds, Jr., and Jay R. Traver, "The Flight Mechanics and Evolution of the Wings of Ephemeroptera, with Notes on the Archetype Insect Wing," *Journal of the Washington Academy of Sciences*, XLIV (1954), 390–400.

R. J. Tillyard has written *The Biology of Dragonflies* (Cambridge, Eng., 1917), and systematic treatments useful to students of the North American fauna are Needham and Minter J. Westfall, Jr., *A manual of the Dragonflies of North America* (Berkeley, Calif., 1955)—does not include the damselflies, suborder Zygoptera—and Edmund M. Walker, *The Odonata of Canada and Alaska* (2 vols.; Toronto, 1953–58).

The dramatic story of early entomological exploration on the Hawaiian Islands (the peculiar terrestrial damselfly which occasioned the discussion of the Hawaiian insect fauna was, however, not discovered until much later) is told by R. C. L. Perkins in *Fauna Hawaiiensis*, Vol. 1, Part VI (London, 1913). A modern account of this fauna, of great interest to the student of evolution, is that of Elwood C. Zimmerman, *Insects of Hawaii: Volume I, Introduction* (Honolulu, 1948).

17. *Beetles*

One of the most useful systematic works dealing with North American beetles has been W. S. Blatchley, *Illustrated and Descriptive Catalogue of the Coleoptera or Beetles of Indiana* (Indianapolis, 1910); a companion volume, by Blatchley and C. W. Leng, *Rhynchophora or Weevils of Northeastern North America* (Indianapolis, 1916), completes the treatment of the order. More modern handbooks that, although not at all giving a complete description of the vast beetle fauna of North America, are essential as introductions to it: Elizabeth and Lawrence S. Dillon, *A Manual of Common Beetles of Eastern North America* (Evanston, Ill., 1961), and Ross H. Arnett, Jr., *The Beetles of the United States* (Washington, D. C., 1960)—this is a manual for the identification of the genera. Bioluminescence, which makes famous the beetles of the family Lampyridae, is discussed by E. Newton Harvey, *Bioluminescence* (New York, 1952).

A general introduction to the Strepsiptera will be found in R. M. Bohart, "A Revision of the Strepsiptera with Special Reference to the Species of North America," *University of California Publications in Entomology*, VII (1941), 91–160. Recent and important information on the biology of a species that parasitizes bees is given by E. G. Linsley and J. W. McSwain, "Observations on the Habits of *Stylops pacifica* Bohart," *University of California Publications in Entomology*, XI (1957), 395–430.

18. *Butterflies and moths*

So long have the butterflies been favorites of collectors that there exist a host of identification guides to these beautiful insects. A standby for the North American fauna has been W. J. Holland, *The Butterfly Book* (2nd ed.; New York, 1946). Two recent books that take long strides toward getting away from the old-fashioned stamp collector's approach and organizing the subject along biological lines are A. B. Klots, *A Field Guide to the Butterflies* (Boston, 1951)—actually includes those of the eastern states only; and Paul R. and Anne H. Ehrlich, *How to Know the Butterflies* (Dubuque, Iowa, 1961). The "macrolepidoptera" (butterflies and most of the families of larger moths) are illustrated and described by A. Seitz, *Macrolepidoptera of the World* (Stuttgart, 1906+); the North and South American butterflies are taken up in Volume 5 (1924).

Among the moths, the multitude of small forms, most of which are included in a more or less arbitrary group called the "microlepidoptera," are very difficult to prepare and study, with the result that such popular works as Holland's *Moth Book* (New York, 1905) more or less ignore them. A technical work that does include them is that of William T. Forbes, *The Lepidoptera of New York and Neighboring States* (New York Agricultural Experiment Station Memoirs, 68; Ithaca, N. Y., 1923). E. B. Ford's *Moths* (New York, 1955), although concerned with the British moths, is well worth the attention of the American reader.

Melanism in moths, of outstanding interest to students of evolution, is the subject of many papers; among them may be mentioned E. B. Ford, "The Study of Organic Evolution by Observation and Experiment," *Endeavour*, XV (1956), 149–52, and H. B. D. Kettlewell, "The Phenomenon of Industrial Melanism in Lepidoptera," *Annual Review of Entomology*, VI (1961), 245–62. Polymorphism in butterflies is among the subjects discussed by Ford, "Polymorphism and Taxonomy," in Julian Huxley, *The New Systematics* (Oxford, 1940), pp. 493–513.

The literature on the biologies of the caterpillars of North American butterflies has been brought together in a continuing series of bibliographic works: Henry Edwards, *Bibliographic Catalogue of the Described Transformations of North American Lepidoptera* (Bulletin of the U. S. National Museum No. 35; Washington, D. C., 1889); Demarest Davenport and V. G. Dethier, "Bibliography of the Described Life Histories of the Rhopolocera of America North of Mexico, 1889–1937," *Entomologia Americana*, XVII, New Series (1938), No. 4; and Dethier, "Supplement to the Bibliography of Described Life Histories of the Rhopolocera of America North of Mexico," *Psyche*, LIII (1946), Nos. 1–2.

19. *The two-winged flies*

S. W. Williston, *Manual of the Families and Genera of North American Diptera* (3rd ed.; New Haven, 1908), and C. H. Curran, *The Families and Genera of North American Diptera* (New York, 1934), are for the most part technical keys for the identification of genera. A guide to the British Diptera, in which one will find excellent illustrations and descriptions of many of the important groups of North American flies, is Charles N. Colyer and C. O. Hammond, *Flies of the British Isles* (London, 1951).

A general account of the mosquitoes is that of Marston Bates, *The Natural History of Mosquitoes* (New York, 1949).

The Siphonaptera are dealt with by Irving Fox and H. E. Ewing, *The Fleas of North America* (United States Department of Agriculture, Miscellaneous Publications No. 500; Washington, D. C., 1943), and G. P. Holland, *The Siphonaptera of Canada* (Canada Department of Agriculture Publication 817, Technical Bulletin 70; Ottawa, 1949).

20. *On stings and societies*

There is in English no adequate book-length account of the Hymenoptera. C. P. Clausen, *Entomophagous Insects* (New York, 1940), has a wealth of information on a wide range of Hymenoptera, since many of these insects are parasites of or predators on other insects, and the social groups are taken up in W. M. Wheeler, *The Social Insects, Their Origin and Evolution* (New York, 1928). The literature on ants is extensive; three books may be referred to here: Wheeler, *Ants, Their Structure, Development, and Behavior*

(New York, 1910); Wilhelm Goetsch, *The Ant* (Ann Arbor, Mich., 1957), and W. S. Creighton, *The Ants of North America* (Harvard Museum of Comparative Zoology, Bulletin 104; Cambridge, Mass., 1950). The last of these is a technical guide to the identification of the species. The Apoidea —in particular the aspects which bear on their classification—have been dealt with by C. D. Michener, "Comparative External Morphology, Phylogeny, and a Classification of the Bees," *American Museum of Natural History Bulletin*, LXXXII (1944), 151–326. Identification of species of bees is facilitated by Theodore B. Mitchell, *Bees of the Eastern United States* (2 vols.; Raleigh, N. C., 1960–62). Two useful sources of information on the domestic honeybee are K. K. Clark, *Beekeeping* (Penguin paperback; Harmondsworth, 1951), and C. R. Ribbands, *Behaviour and Social Life of Honeybees* (London, 1953).

The Hymenoptera of America North of Mexico (Synoptic Catalogue), by C. F. W. Muesebeck, Karl V. Krombein, and Henry K. Townes (Washington, 1951), lists all the species, their ranges, and the major references in the technical literature.

References on social insects cited in the bibliography of Chapter 14, which includes termites, are devoted primarily to Hymenoptera.

Index